# Earth Science Curriculum Activities Kit

Robert G. Hoehn

Illustrations by Chris "Art Director" Reynolds

THE CENTER FOR APPLIED RESEARCH IN EDUCATION
West Nyack, New York 10995

10 9 8 7 6 5 4 3

**Library of Congress Cataloging-in-Publication Data**

Hoehn, Robert G., 1937–
Earth science curriculum activities kit / Robert G. Hoehn.
p. cm.
Includes bibliographical references (p. ).
ISBN 0-87628-288-5
1. Earth sciences—Study and teaching (Secondary) 2. Geophysics—Study and teaching (Secondary) I. Center for Applied Research in Education. II. Title.
QE40.H64 1991
550′.712—dc20 91-2744
CIP

ISBN 0-87628-288-5

**The Center for Applied Research in Education**
Business Information & Publishing Division
West Nyack, NY 10995
Simon & Schuster, A Paramount Communications Company

Printed in the United States of America

# ABOUT THE AUTHOR

ROBERT G. HOEHN has taught earth science, physical science and biology in the Roseville Union School District of California since 1963. He has received six National Summer Science grants from the National Science Foundation and has given numerous presentations to teachers and administrators attending local and state science conventions, workshops, and seminars. He has also served as a mentor teacher in his district.

Mr. Hoehn has a B.A. from San Jose State University and an administration credential from California State University, Sacramento. He is a member of the National Education Association, California Teacher's Association and California Science Teacher's Association. He has published a number of nonfiction books and copymaster sets and over 70 magazine articles on science education, coaching, and baseball/softball strategy.

# ABOUT THIS KIT

*Earth Science Curriculum Activities Kit* offers earth science teachers 126 ready-to-use activities, including laboratory experiments, for grades 7 through 12. The twelve major earth science topics covered are:

- Matter
- Energy Resources
- Minerals
- Rocks
- Volcanoes
- Earthquakes
- Fossils
- Geologic Time Scale
- Forces That Shape the Earth's Surface
- Weather and Climate
- Astronomy
- Oceanography

Each section includes

- Challenging puzzles (word searches, crossword puzzles, and find the missing terms).
- Laboratory experiments and activities that encourage students to make clear, concise observations.
- Activities using easily obtainable, hands-on materials commonly found in most classrooms.
- Brief mini-activities to direct student thinking toward science.

*Earth Science Curriculum Activities Kit* is designed to help students develop an understanding of basic concepts in earth science. Projects in this kit range from matching exercises and puzzles to hands-on laboratory activities. The material, therefore, is useful for supplemental activities during class, for outside assignments, and for extra-credit projects. Many exercises can be done by individuals or small groups of students, and almost all will serve as starting points for class discussion.

Each activity can easily be copied and made available for every student. The classroom laboratory activity lessons follow a simple format: Introduction, Objective, Materials, Procedure, and Questions (when applicable).

Before using any lesson from *Earth Science Curriculum Activities Kit,* preview it to determine its purpose and how much research it requires of your students. You will find that the activities blend well with almost any earth science resource or textbook.

Each section ends with a list of mini-activities. You can give students this list at the beginning of the period or during the last five minutes of the period. A mini-activity, lasting from one to five minutes, can provide a step into the day's lesson or bring the day's work to closure.

Special features of this book are the Teacher's Guide and Answer Key found at the end of each section. The Teacher's Guide provides suggestions, recommendations, and tips for getting the most out of the activities.

*Earth Science Curriculum Activities Kit* will supply you with scores of activities to encourage critical thinking skills and stimulate learning. Enjoy!

*Robert G. Hoehn*

# CONTENTS

About This Kit • v

Section 1 MATTER • 1

1–1 Where's the Matter? (*classroom activity*) 2
1–2 Matter Term Puzzle (*classroom activity*) 3
1–3 Matter Trivia (*classroom activity*) 4
1–4 Arranging Matter (*classroom activity*) 6
1–5 Elements, Elements, Elements (*classroom activity*) 7
1–6 Matter Scatter Puzzle (*classroom activity*) 8
1–7 Matter as a Mixture (*lab activity*) 9
1–8 Another Look at a Mixture (*lab activity*) 10
1–9 Matter as a Compound (*lab activity*) 11

Teacher's Guide and Answer Key • 13

Mini-Activities • 16

Section 2 ENERGY RESOURCES • 19

2–1 Energy Resources Word Search (*classroom activity*) 20
2–2 A Nonrenewable Energy Source: Coal (*classroom activity*) 21
2–3 Hydrocarbon Puzzle (*classroom activity*) 22
2–4 Energy Source Puzzle (*classroom activity*) 23
2–5 Energy Concern (*classroom activity*) 24
2–6 Fuel of the Future (*classroom activity*) 25
2–7 Duster or Gusher? (*classroom activity*) 26
2–8 Making Hydrocarbon Models (*lab activity*) 29
2–9 Burning Coal (*lab activity*) 30
2–10 Simulated Solar Cooker (*lab activity*) 32

Teacher's Guide and Answer Key • 33

Mini-Activities • 36

Section 3 MINERALS • 39

3–1 Mineral Term Game (*classroom activity*) 40
3–2 A Look at Minerals (*classroom activity*) 41

3–3 Mineral Properties (*classroom activity*) 42
3–4 Mineral Identification (*classroom activity*) 43
3–5 Mineral Word Search (*classroom activity*) 44
3–6 How Are Minerals Important to Us? (*classroom activity*) 45
3–7 Hardness (*lab activity*) 46
3–8 Streak (*lab activity*) 47
3–9 Luster (*lab activity*) 48
3–10 Specific Gravity (*lab activity*) 49
3–11 Find the Mystery Mineral (*lab activity*) 50
3–12 Crystals (*lab activity*) 51
3–13 Microcrystals (*lab activity*) 52

Teacher's Guide and Answer Key • 53

Mini-Activities • 56

Section 4 ROCKS • 59

4–1 Sedimentary Rock Puzzle (*classroom activity*) 60
4–2 Metamorphic Rock Puzzle (*classroom activity*) 61
4–3 Igneous Rock Puzzle (*classroom activity*) 62
4–4 Rock Word Search (*classroom activity*) 63
4–5 Matching Rock Properties (*classroom activity*) 64
4–6 Rhyming Rock Puzzle (*classroom activity*) 65
4–7 Sedimentary Rock Observations (*lab activity*) 66
4–8 Igneous Rock Observation (*lab activity*) 68
4–9 Metamorphic Rock Observation (*lab activity*) 70
4–10 Determining the Density of a Rock (*lab activity*) 72

Teacher's Guide and Answer Key • 74

Mini-Activities • 77

Section 5 VOLCANOES • 79

5–1 Volcano Vocabulary Puzzle (*classroom activity*) 80
5–2 Fill-in Mystery Message (*classroom activity*) 83
5–3 Anatomy of a Volcano (*classroom activity*) 84
5–4 The L-term Volcanic Puzzle (*classroom activity*) 85
5–5 Volcanic Rock Observation (*lab activity*) 86
5–6 Specific Gravity of Volcanic Rocks (*lab activity*) 87
5–7 Build a Model Volcano (*lab activity*) 89
5–8 Simulated Volcanic Rock Petroglyphs (*lab activity*) 91

Teacher's Guide and Answer Key • 93

Mini-Activities • 96

## Section 6 EARTHQUAKES • 99

6–1 Earthquake Terms (*classroom activity*) 100
6–2 Earthquakes Everywhere! (*classroom activity*) 101
6–3 Earthquake Word Search (*classroom activity*) 103
6–4 Think About It (*classroom activity*) 104
6–5 Movement Along a Fault (*lab activity*) 105
6–6 Moving Blocks (*lab activity*) 107
6–7 Making Waves (*lab activity*) 108
6–8 Locating the Epicenter (*lab activity*) 110

Teacher's Guide and Answer Key • 114

Mini-Activities • 117

## Section 7 FOSSILS • 119

7–1 What Is a Fossil? (*classroom activity*) 120
7–2 Fossil Term Roundup (*classroom activity*) 122
7–3 Who Was That Dinosaur? (*classroom activity*) 124
7–4 Fossil Clues (*classroom activity*) 125
7–5 A Survey of Bones (*classroom activity*) 126
7–6 Plaster Fossil Model (*lab activity*) 127
7–7 Chalk Stick Fossil Model (*lab activity*) 129
7–8 Wax Imprints (*lab activity*) 130
7–9 Molds and Casts (*lab activity*) 131
7–10 Snail Cast (*lab activity*) 133
7–11 Pencil Print (*lab activity*) 134
7–12 Forever Amber, Part 1 (*lab activity*) 135
7–13 Forever Amber, Part 2 (*lab activity*) 136
7–14 "Reconstruct," a Fossil Game (*game*) 137

Teacher's Guide and Answer Key • 138

Mini-Activities • 143

## Section 8 GEOLOGIC TIME SCALE • 147

8–1 Geologic Time Scale Vocabulary Puzzle (*classroom activity*) 148
8–2 Geologic Time Scale Game (*game*) 150
8–3 How Much Do You Know About the Geologic Timetable? (*classroom activity*) 151
8–4 Examining the Geologic Timetable (*classroom activity*) 152
8–5 Rock Music Timetable (*classroom activity*) 154
8–6 Paleozoic Life Scramble (*classroom activity*) 156
8–7 Mesozoic Life Scramble (*classroom activity*) 157

8–8 Cenozoic Era Word Search (*classroom activity*) 158
8–9 Four Sides: A Geologic Timetable Game (*game*) 159
8–10 Life: Past and Present (*lab activity*) 160

Teacher's Guide and Answer Key • 162

Mini-Activities • 166

**Section 9 FORCES THAT SHAPE THE EARTH'S SURFACE • 169**

9–1 Restless Earth Term Puzzle (*classroom activity*) 170
9–2 Erosion at Work (*classroom activity*) 172
9–3 Weathering Processes (*classroom activity*) 174
9–4 Forces Around Us, Part 1 (*classroom activity*) 176
9–5 Forces Around Us, Part 2 (*classroom activity*) 178
9–6 Glaciers on the Move (*classroom activity*) 180
9–7 Sleuthing the Newspaper (*classroom activity*) 182
9–8 Changing Earth Word Search Puzzle (*classroom activity*) 183
9–9 Diastrophic Events (*classroom activity*) 184
9–10 Pancake Earth (*lab activity*) 185

Teacher's Guide and Answer Key • 190

Mini-Activities • 194

**Section 10 WEATHER AND CLIMATE • 197**

10–1 Weather Word Search (*classroom activity*) 198
10–2 Mixed-up Weather Term Puzzle (*classroom activity*) 199
10–3 Weather Terms Coded Message (*classroom activity*) 200
10–4 Tools of the Meteorologist (*classroom activity*) 202
10–5 Fronts (*classroom activity*) 203
10–6 Weather Trivia (*classroom activity*) 204
10–7 Climate Vocabulary Fill-ins (*classroom activity*) 206
10–8 Water Thermometer, Part 1 (*lab activity*) 208
10–9 Water Thermometer, Part 2 (*lab activity*) 210
10–10 Measuring Humidity with Hair (*lab activity*) 211
10–11 A Straw Barometer (*lab activity*) 213
10–12 Making a Wet- and Dry-Bulb Thermometer (*lab activity*) 214

Teacher's Guide and Answer Key • 216

Mini-Activities • 219

**Section 11 ASTRONOMY • 221**

11–1 Astronomy Word Search (*classroom activity*) 222
11–2 Moon Term Game (*game*) 223

11–3 Nebular Theory (*classroom activity*) 224
11–4 Planet Match-up (*classroom activity*) 226
11–5 What Goes with What? (*classroom activity*) 228
11–6 How Do Planets Compare in Size (Diameter) with the Sun? (*lab activity*) 230
11–7 How About the Moon? (*lab activity*) 232
11–8 Constellations in a Minute! (*lab activity*) 234
11–9 Investigating a Mystery Planet (*lab activity*) 236
11–10 Meteorite Impact (*lab activity*) 238

Teacher's Guide and Answer Key • 240

Mini-Activities • 243

**Section 12 OCEANOGRAPHY • 245**

12–1 Oceanography Vocabulary Puzzle (*classroom activity*) 246
12–2 Things in the Sea (*classroom activity*) 248
12–3 Food Chain (*classroom activity*) 249
12–4 Complete the Term (*classroom activity*) 251
12–5 Secret Message (*classroom activity*) 252
12–6 Instant Sea Water (*lab activity*) 253
12–7 Which Salt Solution Comes the Closest to 3.5% (*lab activity*) 254
12–8 Depth Recording the Old-Fashioned Way (*lab activity*) 255
12–9 What Is the Mystery Object? (*lab activity*) 256
12–10 Miniature Model of a Density Current (*lab activity*) 257
12–11 Can You Beat El Niño? (*game*) 259
12–12 Time to Dine (*game*) 261

Teacher's Guide and Answer Key • 262

Mini-Activities • 265

**BIBLIOGRAPHY • 269**

Section 1

# MATTER

## OUTLINE

1-1 Where's the Matter?
1-2 Matter Term Puzzle
1-3 Matter Trivia
1-4 Arranging Matter
1-5 Elements, Elements, Elements
1-6 Matter Scatter Puzzle
1-7 Matter as a Mixture
1-8 Another Look at a Mixture
1-9 Matter as a Compound

## MATERIAL

The following laboratory materials are needed for Section 1:

- beakers (100 ml)
- beaker tongs
- hammers
- heat sources (Bunsen burners)
- magnets
- matches
- paper tissues
- paper towels
- plastic spoons
- powdered iron
- powdered sulfur
- salt
- sand
- test tube holders
- test tubes
- water

Name ______________________ Date ______________________

# 1-1 Where's the Matter?

Matter is anything that takes up space and has mass. Mass may be defined as a quantity or amount of matter in a body. Therefore, anything a person sees, touches, smells, or tastes is matter.

Identifying matter, then, should be easy. But is it? Let's find out. Use the three hints or clues after each number below to identify the example of matter that can be found on most school campuses.

1. red, white, and blue ______________________
2. white, gypsum, brittle ______________________
3. hydrogen, oxygen, compound ______________________
4. round, countries, third planet ______________________
5. rubber, bounces, "pigskin" ______________________
6. redemption, crumble, metal ______________________
7. turn, metal, lock ______________________
8. record, thin, wood product ______________________
9. friction, rubber, rub ______________________
10. graphite, wood, sharpener ______________________
11. transparent, sand, building ______________________
12. keys, ribbon, tab set ______________________
13. dirt, lanes, hurdles ______________________
14. chapters, words, glossary ______________________
15. wire, conversation, call ______________________
16. learning, human, homework ______________________
17. ending, passing, beginning ______________________
18. leader, supervisor, grader ______________________
19. bends, holds, attaches ______________________
20. hands, numbers, hours ______________________

Name ______________________________ Date ______________________

# 1-2 Matter Term Puzzle

There are 22 terms that are related to matter listed below the puzzle. Nineteen of the terms can be found in the puzzle. Nine of them form a chemical symbol that represents a state of matter without a definite volume or shape. (*Hint:* This state of matter is vital to human life.) The symbol will take shape if you draw a line through the correct nine terms. (*Hint:* As you find the terms, make light lines through the ones that are connected. Then you can erase the ones that seem to go nowhere.) The black boxes may be used as connecting links. Terms may be spelled forward, backward, vertically, horizontally, and diagonally.

|  | A |  | F | C | R | E | T | T | A | M | N |  |
|---|---|---|---|---|---|---|---|---|---|---|---|---|
| C | O | M | P | O | U | N | D |  | E | Z | O | D |
|  | I | A | S | E | G | V | E | Y | P |  | R |  |
| E | A | S | B | R | O |  | L | T | A | E | T | H |
| L |  | S | A | L | I | Y | U | R | Q | B | U | W |
| E | A | M | U | A | S |  | C | E |  | I | E | A |
| C | S | M | S | O | E | G | E | P | J | A | N | P |
| T | E | T | L |  | R | I | L | O | G | A | S | R |
| R | V | I | A | E | K | X | O | R | A |  | U | O |
| O | D |  | N | A | T | O | M | P | T | O | N | T |
| N | L | I | Q | U | I | D |  | A | I | M | L | O |
|  | E | L | E | M | E | N | T |  | N | E | O | N |

atom
centimeter
compound
density
electron
element
gas
grams
liquid
mass
matter
milliliter
molecule
neon
neutron
property
proton
solid
sun
tin
ton
volume

The chemical symbol is ______________. The state of matter is a ______________________

known as ______________________.

Name ______________________________ Date ______________________

# 1-3 Matter Trivia

How good are you at trivia? Find out by answering the following trivia items related to matter. Write your answers in the space provided.

1. The German zeppelin Hindenburg crashed and burned in 1937. Name the combustible matter that caused so much destruction. ______________________________

2. Give an example of how a gas can turn a solid into a gas. ____________________
______________________________________________________________

3. Why is air considered matter if you can't see it? ____________________
______________________________________________________________

4. Name the element that completes this statement: "All that glitters is not ____________________."

5. Name a compound that demonstrates the three phases of matter. ____________________

6. Name three solids that will allow light to travel through them. ____________________, ____________________, and ____________________

7. What round object goes through several phases and never seems to change? __________
______________________________________________________________

8. Matter A and B are the same size. Matter B is twice as heavy as Matter A. Why?
______________________________________________________________

9. What phase of matter does this illustration represent?

______________________________________________________________

10. Why doesn't the following illustration represent a solid?

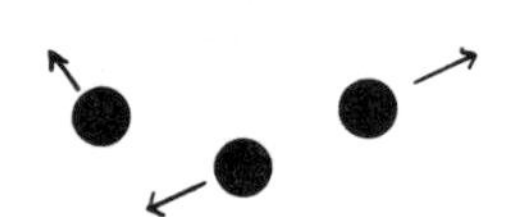

______________________________________________________________

11. What is definite about a liquid? ____________________

12. How can matter be changed from one state to another? ____________________

____________________

13. What do you get when several atoms of the same element join together? ____________

____________________

14. What do you call a substance that contains only one kind of atom? ____________

15. Who was the English scientist (1766–1844) who believed that atoms exist and combine to form matter? ____________________

16. Symbols for elements combine to form chemical ____________ for compounds.

17. When matter exists as a gas, it is referred to as a state or ____________.

18. When a substance changes from a solid to a gas is it still considered matter? Why or why not? ____________________

____________________

19. Complete the following statement regarding a well-known heavy element: "Get the ____________ out."

20. What happens to matter when it changes state? ____________________

____________________



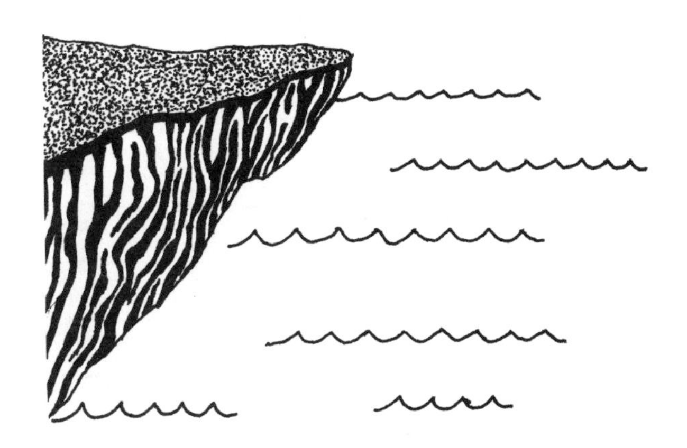

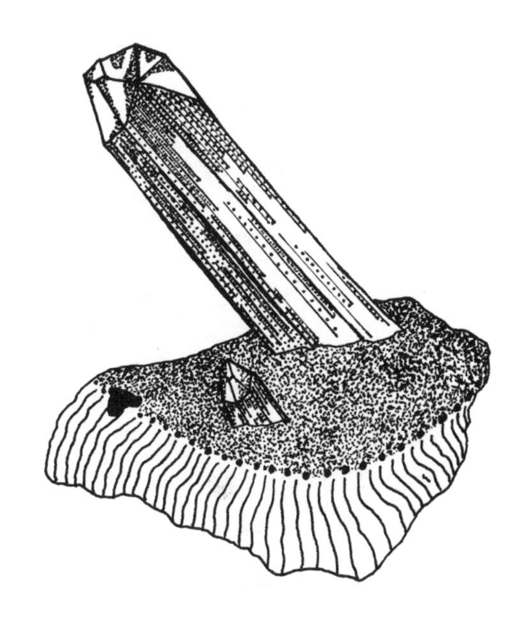

Name ______________________________ Date ______________________

# 1-4 Arranging Matter

Here are 12 statements regarding matter. The statements are jumbled and make little sense in their present form. Unscramble and rewrite each statement in the space provided. Then use the underlined word(s) from each statement to write a short paragraph on matter.

1. quantity of is Mass or matter the amount.

   ______________________________

2. change does mass The not of object an.

   ______________________________

3. The describes matter "stuff" too word.

   ______________________________

4. an Oxygen odorless is tasteless and a gas colorless.

   ______________________________

5. by as matter volume known The occupied space is.

   ______________________________

6. of gas fuel a matter known state produces as Burning.

   ______________________________

7. matter There of are states three different.

   ______________________________

8. shapes not definite have do Liquids.

   ______________________________

9. volume definite shape has A and solid a definite.

   ______________________________

10. property Texture matter a of is.

    ______________________________

11. molecules involve changes in changes Chemical.

    ______________________________

12. matter form An is of element a.

    ______________________________

Paragraph on matter: ______________________________

______________________________

______________________________

______________________________

Name ______________________________ Date ____________________

# 1-5 Elements, Elements, Elements

An element is a basic unit of matter that cannot be changed into any simpler form of matter by ordinary means. Therefore, if you had a 2-ounce hunk of lead and cut off a piece of it, you'd still have lead, but two pieces of different size.

## PART 1

The list below contains 15 different names of elements. Unscramble each name, write the name in the space provided, and shade the squares of each word on the puzzle. The resulting shaded figure will reveal an object familiar to many hikers. Be careful. There are 20 elements in the puzzle; five of these words have no bearing on your results. The words may be found vertically and horizontally.

robanc ______________________

imosud ______________________

utimlhi ______________________

ecnhilro ______________________

boonr ______________________

cisnoil ______________________

oynrghed ______________________

yrrmeuc ______________________

cimlauc ______________________

ynoexg ______________________

sflruu ______________________

elmihu ______________________

orecpp ______________________

rnaog ______________________

cnkeil ______________________



| m | a | g | n | e | s | i | u | m | i | c | f | o | r | s | a |
|---|---|---|---|---|---|---|---|---|---|---|---|---|---|---|---|
| p | o | t | a | s | s | i | u | m | g | a | u | e | k | b | g |
| t | x | o | x | y | g | e | n | a | z | i | n | c | a | e | n |
| l | n |  | c | a | r | b | o | n | l | e | a | d | p | m | i |
| s | i | c | h | s | o | d | i | u | m | c | o | b | a | l | t |
| u | c | o | e | c | a | l | c | i | u | m | l | f | o | n | j |
| l | k | p | l | a | l | i | t | h | i | u | m | h | e | l | r |
| f | e | p | i | r |  | m | e | r | c | u | r | y | a | d | a |
| u | l | e | u | g |  |  | s | i | l | i | c | o | n | i | m |
| r |  | r | m | o |  |  | h | y | d | r | o | g | e | n | t |
| b | o | r | o | n |  |  |  | c | h | l | o | r | i | n | e |

The shaded pattern reveals the profile of a ______________

______________________.

## PART 2

Now list the 15 elements in alphabetical order on the back of this sheet. Write the chemical symbol for each element next to the element.

Name ______________________________ Date ______________________

# 1-6 Matter Scatter Puzzle

## PART 1

Locate and circle the 15 examples of matter in the puzzle. Some are living organisms or structures; others are nonliving items. Use the Hint List to help you find the words. The words may be found forward, backward, vertically, horizontally, or diagonally. The same letters may be used in more than one word, so to make things easier, write the words in a list as you locate them so you can arrange them in alphabetical order on the back of this sheet.

**Hints**

1. A gas found everywhere on Earth
2. A pollinating insect
3. A bed-like structure
4. Carbon dioxide, for example
5. Part of the foot
6. Used to tie things together
7. Anchors a plant
8. A domestic animal
9. A sealed container
10. Growing out of the skin
11. A cutting tool
12. Found at the beach
13. Mode of transportation
14. A bone found in the chest area
15. A cobra or moccasin, for example

| | | | | | | | |
|---|---|---|---|---|---|---|---|
| E | I | Y | C | H | S | N | W |
| K | P | O | A | A | B | F | A |
| A | T | G | N | I | R | T | S |
| N | A | D | R | R | O | O | T |
| S | C | E | E | B | I | E | T |
| T | A | L | W | O | V | G | P |



## PART 2

Complete the following items:

1. Three parts of the human body are listed in the puzzle. They are ________________, ________________, and ________________.

2. Three members of the animal family are listed in the puzzle. They are ________________, ________________, and ________________.

3. The three states of matter are solid, liquid, and gas. The only state of matter not included in the puzzle is ________________.

Name ______________________ Date ______________

# 1-7 Matter as a Mixture

## PART 1

**Introduction:** If two compounds are mixed together, they remain as a mixture unless a chemical change occurs. For example, if sand is mixed with sugar, the two do not combine chemically. The ingredients—sand and sugar—are a mixture of two compounds. A mixture can usually be separated by a simple physical change.

**Objective:** To separate a mixture by a simple physical change.

**Materials:** Salt, sand, water, beaker (100 ml), heat source, beaker tongs, and matches.

**Procedure:** Use any of the listed materials to solve the following problem:

*Set up an experiment which shows that a salt solution added to sand is a mixture, not a chemical change. Use the guidelines below to record experimental results.*

1. Title of experiment:
2. Problem (hypothesis):
3. Materials:
4. Procedure (steps in the experiment):
5. Results:



## PART 2

If you add lettuce, tomatoes, radishes, onions, and peppers together, this mixture produces a salad. If you put stamps from 20 different countries in the same envelope and shake them up, you have a mixture of stamps. On the back of this sheet, list five people (doctor, teacher, butcher, etc.) who make mixtures almost every day. Describe each of the mixtures.

Name ______________________ Date ______________________

# 1-8 Another Look at a Mixture

## PART 1

**Introduction:** Mixtures may be made from any amounts of two or more elements. You can have more or less of any one element; for example, 5 grams of Element A mixed with 13 grams of Element B.

**Objective:** To separate a mixture by a simple physical change.

**Materials:** Powdered sulfur, powdered iron, plastic spoon, magnet, beaker (100 ml), and paper tissue.

**Procedure:** Do the following:

1. Pour a half spoonful of iron powder into a beaker. List three characteristics of iron.

   a. ______________________

   b. ______________________

   c. ______________________

2. Add one spoonful of sulfur powder into the beaker. Do not mix. List three characteristics of sulfur.

   a. ______________________

   b. ______________________

   c. ______________________

3. Mix the two materials together. List three characteristics of the mixture.

   a. ______________________

   b. ______________________

   c. ______________________

4. Cover an end of the magnet with paper tissue. Try to recover as much iron from the mixture as you can. Scrape the iron into a clean spoon.

   Did you recover a half spoonful of iron from the mixture? Why or why not?

   ______________________

   ______________________

## PART 2

Fill in the blands.

The mixture contained mostly ______________ and a small amount of ______________. This mixture contained ______________ amounts of the two elements.

**Name** ______________________________ **Date** ____________________

# 1-9 Matter as a Compound

**Introduction:** The iron/sulfur mixture from Activity 1-8 can be changed into a compound by melting the mixture. If the mixture is heated until it glows red for several minutes, a compound known as ferrous sulfide will form. Does the compound resemble the iron/sulfur mixture? Can these elements be separated? Let's find out.

**Objective:** To make a compound from a mixture.

**Materials:** Iron/sulfur mixture from Activity 1-8, Bunsen burner, test tube, test tube holder, hammer, paper towel, magnet, apron, and safety goggles.

**Procedure:** Do the following:

1. Put the iron/sulfur mixture in a test tube.
2. Heat the mixture until it glows red. Continue heating for several minutes.
3. Allow the mixture to cool. Wrap a paper towel around the test tube. Break the tube with a hammer.
4. Recover the fragments. Pass the magnet over the fragments.

Now answer the following:

1. List three characteristics of the iron/sulfur mixture.

   a. ______________________________

   b. ______________________________

   c. ______________________________

2. Can you recognize sulfur particles? Why or why not? ______________________________

   ______________________________

3. Can you recognize iron particles? Why or why not? ______________________________

   ______________________________

4. Was the magnet attracted to the compound? Why or why not?

   ______________________________

5. Name a property sulfur lost after the compound formed.

   ______________________________

6. Name a property iron lost after the compound formed.

   ______________________________

7. A new substance is produced when a chemical change occurs. You have seen how the original substance, the iron/sulfur mixture, formed different properties. Why do you think this happened?

______________________________________________________________

______________________________________________________________

8. List two ways a compound and a mixture are alike.

   a. ____________________________________________________________

   b. ____________________________________________________________

9. List two ways a compound and a mixture are different.

   a. ____________________________________________________________

   b. ____________________________________________________________



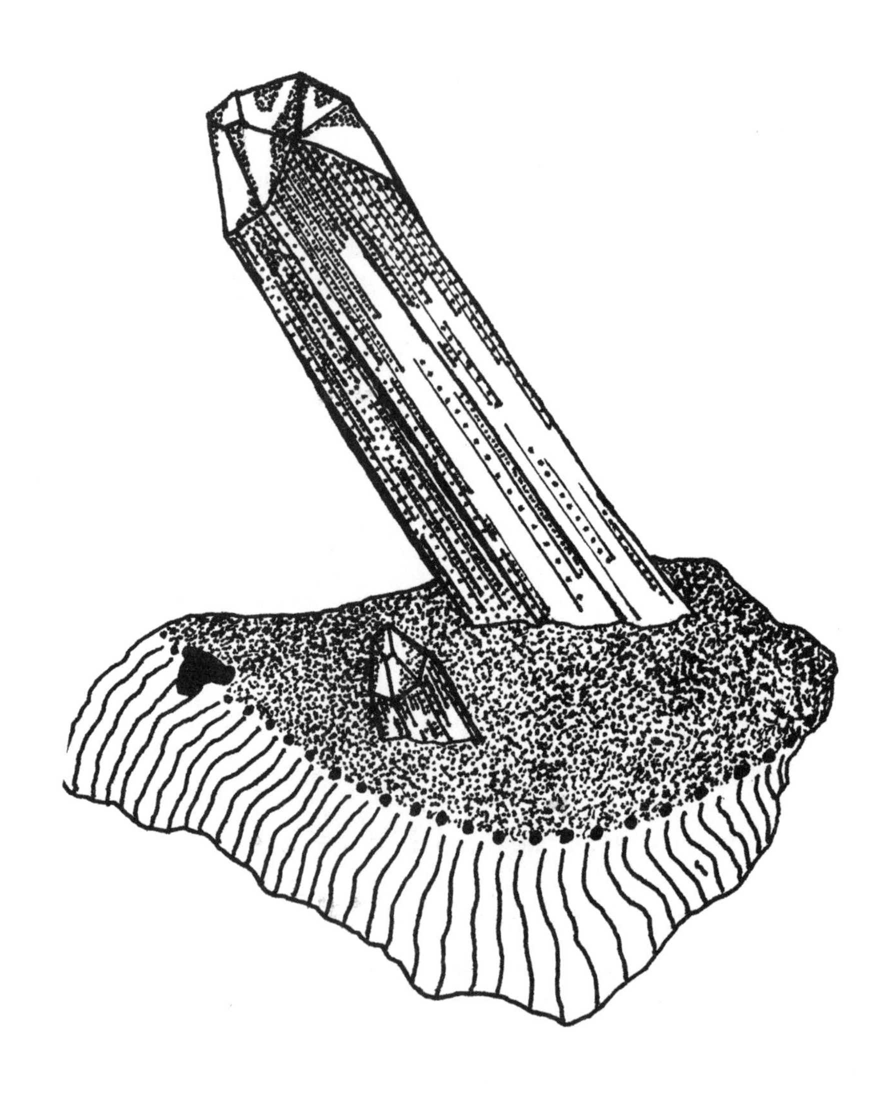

## Section 1: Matter
## TEACHER'S GUIDE AND ANSWER KEY

### 1-1 Where's the Matter?

1. flag; 2. chalk; 3. water; 4. world globe; 5. football; 6. soft drink container; 7. door handle or key; 8. paper; 9. eraser; 10. pencil; 11. window; 12. typewriter; 13. track; 14. textbook; 15. telephone; 16. student; 17. bell; 18. teacher; 19. paperclip; 20. clock

### 1-2 Matter Term Puzzle

| | A | | F | C | R | E | T | T | A | M | N | |
|---|---|---|---|---|---|---|---|---|---|---|---|---|
| C | O | M | P | O | U | N | D | | E | Z | O | D |
| | I | A | S | E | G | V | E | Y | P | | R | |
| E | A | S | B | R | O | | L | T | A | E | T | H |
| L | | S | A | L | I | Y | U | R | Q | B | U | W |
| E | A | M | U | A | S | | C | E | | I | E | A |
| C | S | M | S | O | E | G | E | P | J | A | N | P |
| T | E | T | L | | R | I | L | O | G | A | S | R |
| R | V | I | A | E | K | X | O | R | A | | U | O |
| O | D | | N | A | T | O | M | P | T | O | N | T |
| N | L | I | Q | U | I | D | | A | I | M | L | O |
| | E | L | E | M | E | N | T | | N | E | O | N |

The chemical symbol is $O_2$. The state of matter is a *gas* known as *oxygen.*

### 1-3 Matter Trivia

1. hydrogen; 2. a gas flame burning paper; 3. it takes up space and has mass; 4. gold; 5. water; 6. glass, cellophane, clear plastic; 7. the moon; 8. matter B has a higher density; 9. solid; 10. the molecules are moving apart; 11. its volume; 12. by adding or taking away heat; 13. molecules; 14. element; 15. John Dalton; 16. formulas; 17. phase; 18. yes, it only changes into a different state or phase; 19. lead; 20. it takes a different form, its molecular composition changes.

## 1-4 Arranging Matter

1. Mass is the quantity or amount of matter.
2. The mass of an object does not change.
3. The word "stuff" describes matter too.
4. Oxygen is an odorless, colorless, and tasteless gas.
5. The space occupied by matter is known as volume.
6. Burning fuel produces a state of matter known as gas.
7. There are three different states of matter.
8. Liquids do not have definite shapes.
9. A solid has a definite shape and a definite volume.
10. Texture is a property of matter.
11. Chemical changes involve changes in molecules.
12. An element is a form of matter.

The paragraphs on matter will vary.

## 1-5 Elements, Elements, Elements

If student texts do not include a periodic table of elements, provide charts for students. A large wall chart will work well for this activity.

### *Part 1*

(left column) carbon, sodium, lithium, chlorine, boron, silicon, hydrogen, mercury; (right column) calcium, oxygen, sulfur, helium, copper, argon, nickel

The shaded pattern reveals the profile of a flat-topped hill or mountain.

### *Part 2*

1. argon, Ar; 2. boron, B; 3. calcium, Ca; 4. carbon, C; 5. chlorine, Cl; 6. copper, Cu; 7. helium, He; 8. hydrogen, H; 9. lithium, Li; 10. mercury, Hg; 11. nickel, Ni; 12. oxygen, O; 13. silicon, Si; 14. sodium, Na; 15. sulfur, S

## 1-6 Matter Scatter Puzzle

### *Part 1*

1. air; 2. bee; 3. cot; 4. gas; 5. toe; 6. string; 7. root; 8. cat; 9. can; 10. hair; 11. saw; 12. sand; 13. car; 14. rib; 15. snake

### *Alphabetical Order*

air, bee, can, car, cat, cot, gas, hair, rib, root, sand, saw, snake, string, toe

### *Part 2*

1. hair, rib, toe; 2. bee, cat, snake; 3. liquid

## 1-7 Matter as a Mixture

Give students just enough information to help them start their experiments. Ask them to select a problem and test it. Allow them to try two or three different hypotheses. If they experience prolonged difficulty, have them ask themselves, "How can I separate sand from salt and salt from salt water?"

### *Part 1*

Students' experiments will vary. Most students will evaporate or boil away the water, leaving behind a salt residue. The salt residue may be scraped or rubbed off the sand grains.

### *Part 2*

Answers will vary. Here are some possible responses: 1. pharmacist, mixes several drug ingredients; 2. knitter, mixes various colored threads; 3. gardener, mixes different flower seeds; 4. writer, mixes short and long sentences; 5. artist, mixes paints while drawing

## 1-8 Another Look at a Mixture

### *Part 1*

1. a. black color, b. fairly heavy, c. has properties of a metal, attracts to a magnet; 2. a. yellow color, b. strong odor, c. soft; 3. a. yellowish-black color, b. strong odor, c. the elements mix well together; 4. almost, some of the iron particles stick to the sulfur, it is difficult to recover the original amount.

### *Part 2*

sulfur, iron, varying

## 1-9 Matter as a Compound

Encourage students to heat the mixture over a flame for several minutes. Have them heat the mixture for at least three or four minutes after the mixture turns red. (Inadequate heating fails to demagnetize the iron.) Make sure the room is well ventilated because the heated mixture gives off a strong sulfur odor.

1. a. It is black in color.
   b. The iron atoms have combined with the sulfur atoms.
   c. Fragments are hard, chunky pieces; gives off a sulfur smell resembling rotten eggs.
2. Probably not; they have melted.
3. Probably not; they have fused or melted with the sulfur particles.
4. No. The iron is no longer attracted to the magnet.
5. color, texture
6. Its attraction to a magnet
7. The sulfur atoms and iron atoms have combined to make a different substance—ferrous sulfide.
8. Both require different elements and demonstrate certain characteristics or property features.
9. A compound is the result of a chemical change (that is, a new substance is formed). The new substance cannot be separated by simple physical means. A mixture can be separated by simple physical means. In a compound, the atoms fuse together; in a mixture, the atoms of combined elements do not fuse together.

# Section 1: Matter
## MINI-ACTIVITIES

Listed below are 10 mini-activities, one to five minutes in length, for students to do at the beginning or end of the period.

1. Complete the crossword puzzle.

**Across**

2. Has mass; takes up space
4. Two elements form a ________.
5. Copper and aluminum are examples.

**Down**

1. Metric unit of mass
3. Fe, Si, and C are all examples.

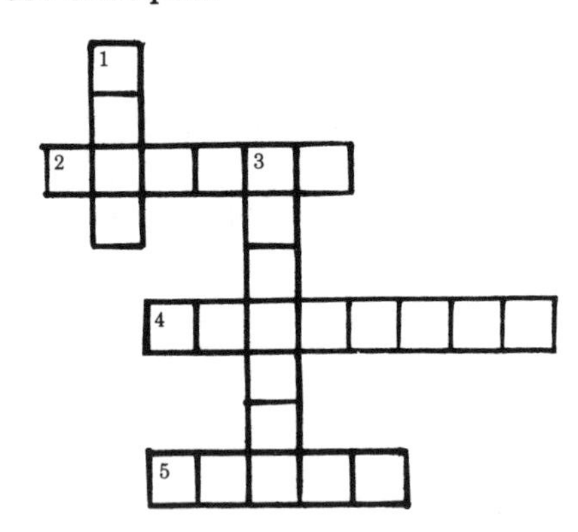

(**Answers:** *Across*—2. matter 4. compound 5. metal *Down*—1. gram 3. element)

2. Chemical symbols represent a shorthand way to write the names of elements. For example, S for sulfur, C for carbon, Al for aluminum. Create your own symbols for the following make-believe elements. Use no more than two letters.

   a. chilloutium
   b. rawdeala
   c. rockanrolium
   d. lightenup
   e. nowayzia
   f. slidenbyzium

   (Answers will vary.)

3. Create six make-believe chemical elements and symbols. Base their names on a hobby, sport, or anything that interests you.
   (Answers will vary.)
4. **Riddle:** You've heard of the element of surprise and the element of chance. What would you call the following two elements?

   Lu    Su

   (**Answers:** Lu = element of luck; Su = element of surprise)
5. **Riddle:** Mr. Fred Matter owns three lots, two houses, and one apartment building. How would you describe what he owns? (**Answer:** The property or properties of matter)
6. **Riddle:** Two nitrogen atoms and one oxygen atom decided to have a party. They spent most of the time sitting around telling jokes. What was the result?
   (**Answer:** laughing gas, $N_2O$)

7. Use the following hints to discover the mystery matter. Then shade the squares containing the letters that make up its chemical formula. *Hints:* a gas; dissolved in many kinds of beverages; found in fire extinguishers

| | | | |
|---|---|---|---|
| n | o | s | h |
| g | a | e | c |
| t | m | r | k |
| i | f | o | b |

(**Answer:** carbon dioxide, C + O + O or $CO_2$)

8. Use the following hints to discover the mystery matter. Then shade the squares containing the letters that make up its chemical formula. *Hints:* a solid; conducts heat and electricity; has a specific luster; may be malleable or sectile

| | | | |
|---|---|---|---|
| e | a | l | b |
| k | i | c | d |
| u | r | m | f |
| w | o | v | g |

(**Answer:** Cu, copper)

9. Use the following hints to discover the mystery matter. Then shade the squares containing the letters that make up its chemical formula. *Hints:* a liquid; an important oxide; necessary for life; boils at 100°C

| | | | |
|---|---|---|---|
| h | a | s | d |
| g | i | e | b |
| l | h | n | k |
| m | c | f | o |

(**Answer:** water, H + H + O or $H_2O$)

10. Combine the following letters and symbols to solve the riddle and answer this question:

*Who was the man that believed all matter was made of four elements?* ____________

____________________

A + + 15th letter + t + found between k and m + last letter in "the"

(**Answer:** A + rist + o + t + l + e = Aristotle)

Section 2

# ENERGY RESOURCES

## OUTLINE

2-1 Energy Resources Word Search
2-2 A Nonrenewable Energy Source: Coal
2-3 Hydrocarbon Puzzle
2-4 Energy Source Puzzle
2-5 Energy Concern
2-6 Fuel of the Future
2-7 Duster or Gusher?
2-8 Making Hydrocarbon Models
2-9 Burning Coal
2-10 Simulated Solar Cooker

## MATERIAL

The following laboratory materials are needed for Section 2:

- alcohol burners
- aluminum foil
- balances
- beaker tongs
- beakers, 50 ml
- candles
- quarters
- forty-watt lamps for light sources
- funnels
- glue
- magnifiers
- matches
- microscope slides
- mirrors
- mortars and pestles
- pens or pencils
- rings and ring stands
- rulers
- safety goggles
- samples of peat, lignite, bitumen, and anthracite
- scissors
- tape
- thermometers
- thread
- various sized beakers
- water
- white paper, 8½ × 11 inch
- wooden splints
- yellow and black construction paper

Name ______________________ Date ______________________

# 2-1 Energy Resources Word Search

Locate and circle the 31 terms related in some way to energy resources. The terms are listed below the puzzle. They may be found backward, forward, vertically, horizontally, and diagonally.

```
k a l c g e n e r a t o r b y n y s
e e i s l e u f l i s s o f t a r a
r l m n o i t s u b m o c u i t a n
o b e s u o n i m u t i b s c u t d
s a s s d j g h i f r r e i i r n s
e e t e n i b r u t a m l o r a e t
n m o k p r t y c e s o u n t l m o
e r n s x y g e l o n z q d c g i n
w e e f h r l c l v e t n b e a d e
c p d i e e u a c o a i e k l s e a
g m o n o n r e n e w a b l e c s b
b i e r g e j l p f c e t i n g i l
h y d r o c a r b o n a m s h a l e
i y w a t e r t a n t h r a c i t e
h q e w c s h l a m r e h t o e g n
r a e l b a e m r e p c i n a g r o
a v i d y n o i s s i f a u c p i x
```

| | | |
|---|---|---|
| anthracite | hydrocarbon | organic |
| bituminous | hydroelectric | peat |
| coal | impermeable | permeable |
| combustion | kerosene | sandstone |
| electricity | lignite | sedimentary |
| energy | limestone | shale |
| fission | natural gas | solar |
| fossil fuels | nonrenewable | turbine |
| fusion | nuclear | water |
| generator | oil | wind |
| geothermal | | |

Name ______________________________ Date ______________________

# 2-2 A Nonrenewable Energy Source: Coal

## PART 1

Unscramble the term related to coal or coal development found in the left-hand column. Write the correct term in the space to the left of the number. Place the number of the unscrambled term in the space to the right of the term.

| | | | |
|---|---|---|---|
| ______________ | 1. "kabcl dlgo" | ____ | a. Coal is this type of rock. |
| ______________ | 2. tsnoimbuui | ____ | b. Partly decayed plant matter |
| ______________ | 3. ncaorb | ____ | c. Containing carbon |
| ______________ | 4. mlciheca | ____ | d. Brown coal |
| ______________ | 5. tpae | ____ | e. Hard coal |
| ______________ | 6. repserus | ____ | f. A change that turns peat into lignite |
| ______________ | 7. psmwa | ____ | g. Responsible for turning soft coal into hard coal: heat and __________ |
| ______________ | 8. pctimhearom | ____ | h. Another name for coal |
| ______________ | 9. ysreadtinem | ____ | i. Substance found in all organic compounds |
| ______________ | 10. grcnioa | ____ | j. Soft coal |
| ______________ | 11. ntreahatic | ____ | k. Original home for plants that eventually turned into coal |
| ______________ | 12. tgnieli | ____ | l. Anthracite is this type of rock. |



## PART 2

The following six statements describe how coal is formed. They need to be placed in the correct order of events. Write the statements, in order, on another sheet of paper.

1. More pressure changes peat into a low-grade coal known as lignite.
2. Plant materials—roots, twigs, leaves, stems—become buried in a swamp environment.
3. More heat and pressure change bituminous coal into anthracite (hard coal).
4. Peat reveals the partial remains of plant material—twigs and leaves.
5. Continued pressure changes lignite into soft coal.
6. Plant material, buried by overlying sediments, undergoes increasing pressure.

Name ____________________ Date ____________________

# 2-3 Hydrocarbon Puzzle

Hydrocarbons are organic compounds that contain hydrogen (H) and carbon (C). They are referred to as fossil fuels. Fossil fuels were formed millions of years ago from organic matter—plants and animals—buried in the Earth's crust. They provide much of the world's energy in the form of coal, oil, and natural gas.

Use the hints below to fill in the blanks of the terms related to the formation of hydrocarbons. Then, on the back of this sheet, briefly tell how each term in the puzzle relates to hydrocarbons.

1. _ _ _ H _ _ _ _ _ _
2. _ Y _ _ _ _ _ _
3. _ _ D _ _ _ _ _ _ _ _
4. _ _ R _ _ _
5. _ _ _ _ O _ _ _ _
6. C _ _ _
7. _ _ A _
8. _ R _ _ _ _ _
9. B _ _ _ _ _ _ _ _ _
10. _ _ _ O _ _ _ _ _
11. _ _ _ N _ _ _
12. S _ _ _ _



### Hints

1. Hard coal; metamorphic rock
2. Light, colorless gas; usually contains one proton, one electron, and no neutrons
3. Rock formed of sediments; a buildup of rocks from wind or water deposits
4. A nonmetallic element found in all organic substances; chemical symbol: C
5. Also called oil; a liquid, flammable substance
6. A black, brittle fossil fuel
7. Partially carbonized plant material
8. A gas obtained from petroleum; chemical formula: $C_3H_8$
9. Another name for soft coal
10. Any natural products that can be drawn on
11. Another name for brown coal
12. A sedimentary rock that changes to slate under extreme heat and pressure

Name ______________________________ Date ____________________

# 2-4 Energy Source Puzzle

Read the clues and use the words in the word list to complete the puzzle. The words all refer to sources of energy.

**Word List**

coal
geothermal
hydroelectric
natural gas
nuclear
petroleum
solar
synfuels
tidal power
wind generator

**Across**

3. propeller and moving air
4. colorless and odorless
5. sun and light
7. black and oily
9. dam and coast
10. synthetic and expensive

**Down**

1. rivers and streams
2. bituminous and anthracite
6. fusion and fission
8. hot water and steam

Name ________________________ Date ________________________

# 2-5 Energy Concern

Many newspaper and magazine articles deal with energy issues such as these: Are we using our energy sources wisely? Are our energy conservation measures inadequate to support future generations? Is our energy resource bank nearly empty? and so on.

Find an article discussing a concern related to energy. Read the article and complete the following items:

1. Title of article: ________________________
2. Name of publication: ________________________
3. List the main concern(s), problem(s), or question(s) brought out in the article. ________

________________________

________________________

________________________

4. Write 10 true statements based on the information in the article.

***For example:***

*Geologists are considering exploring for oil in an unspoiled stretch of Arctic wilderness.*

________________________

________________________

________________________

________________________

________________________

________________________

________________________

________________________

________________________

________________________

________________________

If you were attending a meeting regarding the main issue described in the article, what three questions might you ask? Write these on the back of this sheet.

Name ______________________________ Date ______________________

# 2-6 Fuel of the Future

## PART 1

Write the letter of the correct answer in the blank.

__ 1. Fossil fuels are nonrenewable energy sources. Which of the following is not a fossil fuel? a. petroleum; b. lignite; c. natural gas; d. oxygen

__ 2. Fossil fuels may be replaced in the future by ______. a. electricity; b. nuclear energy; c. oil; d. thermal conduction

__ 3. Oil can be extracted from: a. coal seams; b. tar sands; c. slate; d. geothermal products

__ 4. Oil shales and tar sands of the world are believed to contain ______ more oil than in the remaining oil sources. a. 50%; b. 40%; c. 20%; d. 10%

__ 5. The largest geothermal power plant in the world is located in: a. Montana; b. Idaho; c. Florida; d. California

__ 6. Which of the following is an unconventional source of energy? a. sun; b. nuclear fission; c. gas under great pressure (geopressure); d. ocean

__ 7. Which of the following statements presently describes our energy future? a. Most energy will come from geothermal power; b. Obtaining energy will not be easy and it will be expensive; c. We will depend exclusively on solar energy; d. Nuclear energy will solve 95% of our problems.



## PART 2

Fill in the blanks with the correct answer of the choices in parentheses.

1. The Earth's heat is referred to as ______________ heat. (geothermal, radiant, crustal)

2. Geothermal energy can provide a safe and clean source of fuel to replace ______________ fuels. (shale, fossil, renewable)

3. Hydroelectric power can be used to turn generators to make ______________. (geothermal energy, gasohol, electricity)

4. Any plant or animal matter that is used as a source of energy is called ______________. (biomass fuel, solar fuel, thermorganic fuel)

5. Fuels made from turning coal into a gas or liquid are called ______________. (conventional fuels, liquafuels, synfuels)

## PART 3

Draw a line connecting the term in the left-hand column with the description in the right-hand column.

| Term | Description |
|---|---|
| 1. solar energy | a. heat from the Earth's interior |
| 2. conservation | b. a metamorphosed bituminous product |
| 3. plastics and medicine | c. raw material made from petroleum |
| 4. oil | d. renewable resource |
| 5. fossil fuel | e. the preservation of natural resources for economical use |
| 6. geothermal energy | f. nonrenewable resource |



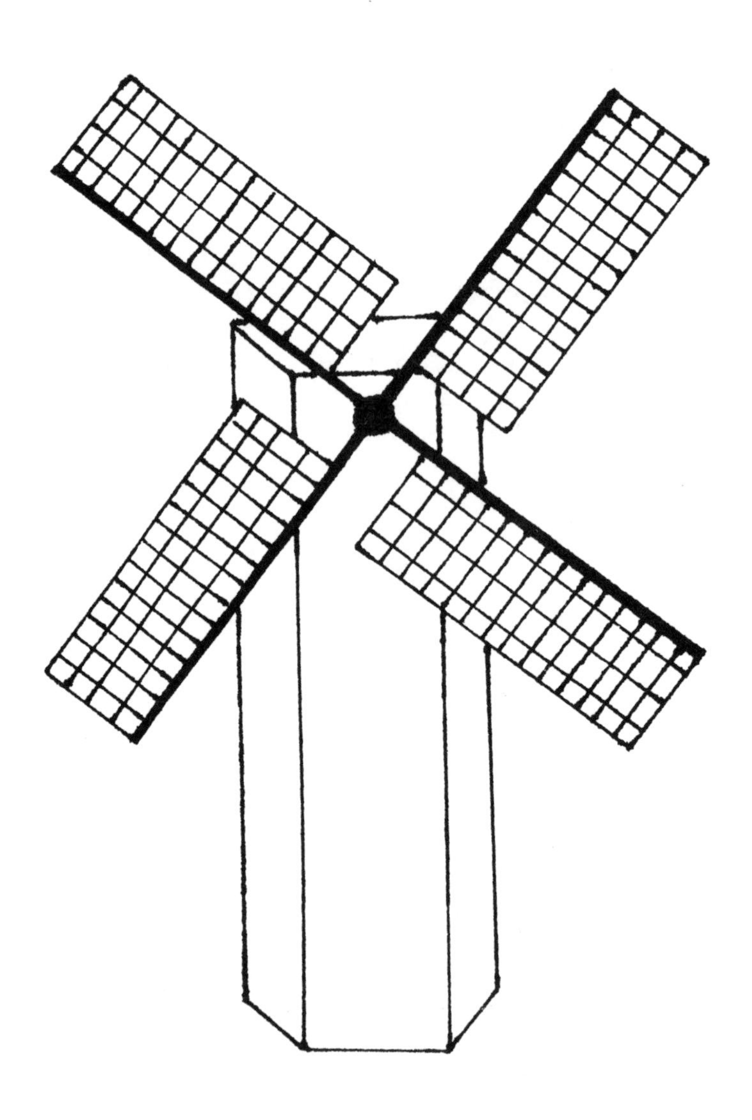

Name ______________________ Date ______________________

# 2-7 Duster or Gusher?

How good are you at finding oil? That's a hard question to answer because you've probably never searched for oil! So here's your chance. Find a partner and play "Duster or Gusher?"

Follow this procedure, using the gameboard to hit it big:

1. Gather these materials: red and white dry beans (to be used as markers) and one die.
2. Play begins when each partner rolls the die. The player rolling the higher number begins first.
3. Each player is allowed one roll of the die per turn. If a marker lands on a letter or symbol appearing on the game sheet, the player checks its meaning and value on the symbol chart. For example, if a player's marker lands on SS (loose sandstone), he or she moves four spaces ahead.
4. The first player who reaches the oil well on the finish line wins the game.

## SYMBOL CHART

| Symbol or Letter | Meaning | Value |
|---|---|---|
| D | Duster—a dry hole | Go back to start. |
| SW | Salt water—oil may be nearby. | Move ahead one space. |
| Sh | Hard shale—nonporous rock | Move back two spaces. |
| Ls | Coarse limestone—porous rock | Move ahead one space. |
| Gr | Granite—igneous, nonporous rock | Move back three spaces. |
| Ba | Basalt—igneous, nonporous rock | Move back three spaces. |
| SS | Loose sandstone—oil-bearing rock | Move ahead four spaces. |
| ▬ | Rest stop; take a break | Lose one turn. |
|  | Foraminifera present in core sample | Move ahead three spaces. |
|  | Broken drill bit; change bits | Move back five spaces. |

## Gameboard

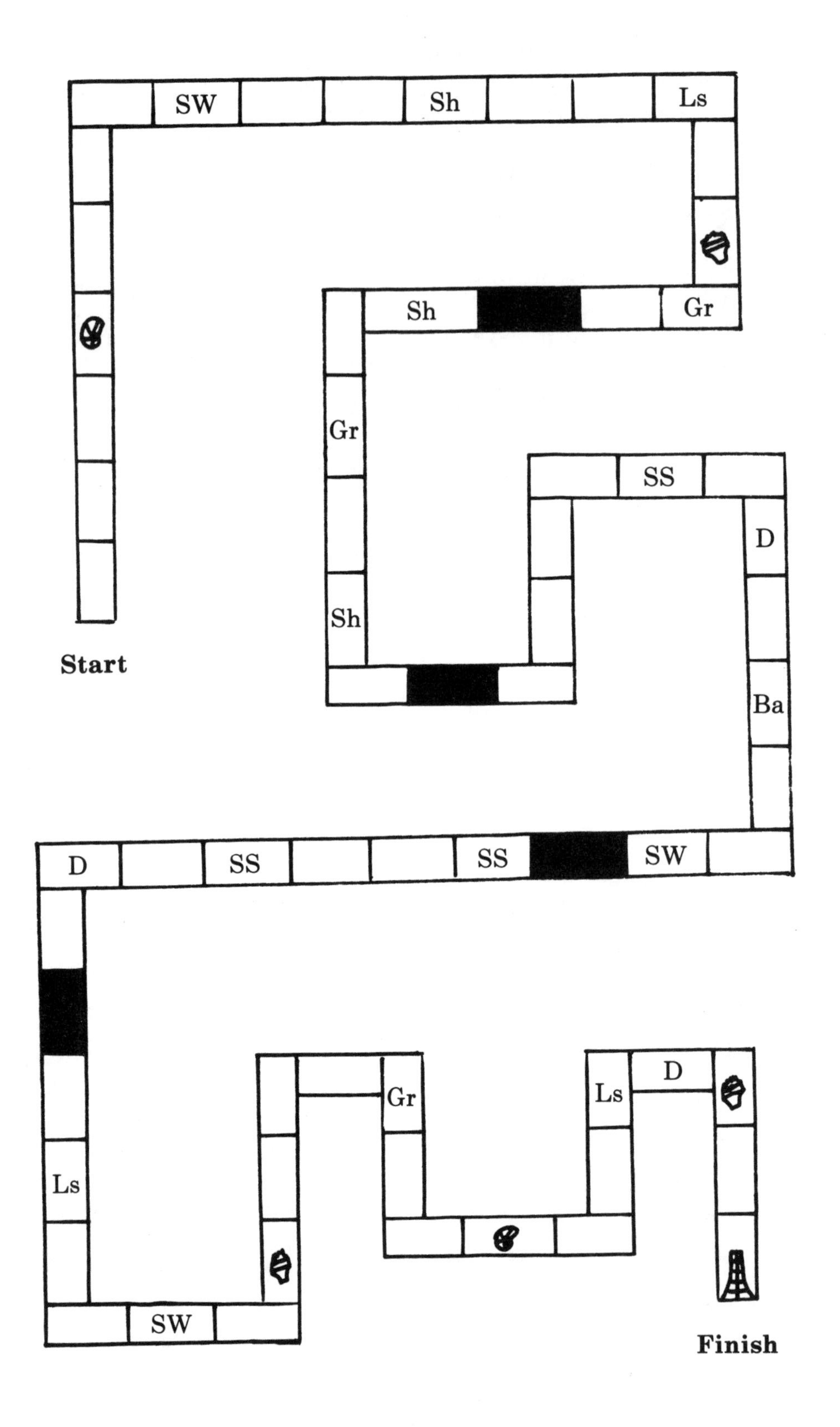

Name ______________________ Date ______________________

# 2-8 Making Hydrocarbon Models

**Introduction:** Carbon and hydrogen combine to form hydrocarbon compounds that make up crude oil. For example, $CH_4$, methane gas, is a hydrocarbon—one carbon atom combined with four hydrogen atoms. $C_8H_{18}$, octane (liquid), is a hydrocarbon—eight carbon atoms combined with 18 hydrogen atoms.

How are hydrocarbons important to the petroleum industry? Hydrocarbons such as gasoline, kerosene, lubricating oil, and fuel oil can be separated from crude oil. People find these products extremely useful.

**Objective:** To make models of hydrocarbon products.

**Materials:** Scissors, glue, a quarter, yellow and black construction paper, white plain paper, ruler, and pen or pencil.



**Procedure:** Do the following:

1. Use the sketches provided to help you build hydrocarbon models. Let black stand for carbon; yellow for hydrogen.
2. Place a quarter on the colored paper and draw a circle by running a pen or pencil along the edge of the quarter.
3. You'll need to make 28 black circles and 70 yellow circles to complete the seven hydrocarbon models. Use scissors to cut out each circle. Construct the models by gluing the colored circles on white paper. Keep circles about one-quarter inch apart. Draw a straight line connecting the circles.
4. Write the formula and name of the hydrocarbon under each model.

**Model Sketches**

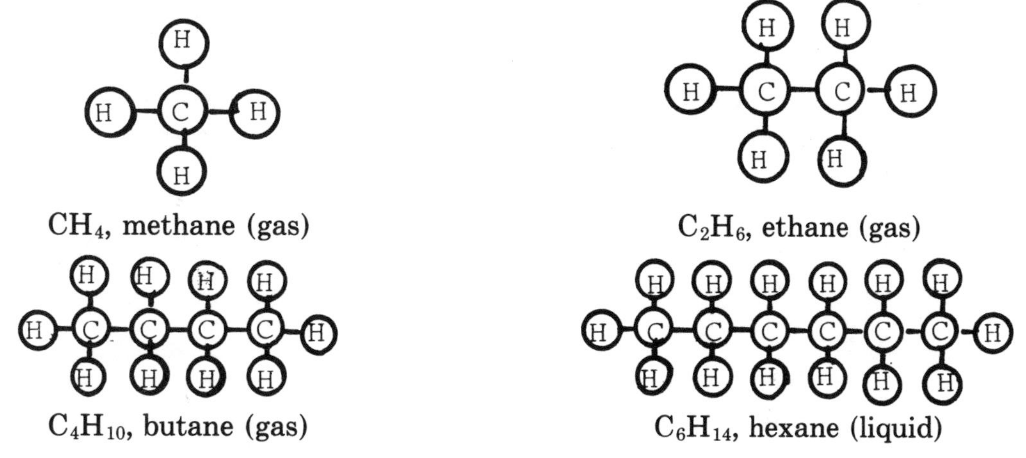


$CH_4$, methane (gas)

$C_2H_6$, ethane (gas)

$C_4H_{10}$, butane (gas)

$C_6H_{14}$, hexane (liquid)

Build the next three models using the illustrations of the model hydrocarbons as guidelines.

1. $C_3H_8$, propane (gas)
2. $C_7H_{16}$, heptane (liquid)
3. $C_5H_{12}$, pentane (liquid)

Name ______________________ Date ______________________

# 2-9 Burning Coal

**Introduction:** When coal burns it releases the trapped energy of trees and plants that grew in swamps millions of years ago. The four stages of coal development are peat, lignite, bituminous (soft) coal, and anthracite (hard coal). Find out how each stage differs by the amount of ash left behind when burned.

**Objective:** To examine the amount of coal ash left after burning.

**Materials:** Samples of peat, lignite, bituminous and anthracite, balance, 50 ml beaker, alcohol burner, mortar and pestle, beaker tongs, ring and ring stand, and safety goggles.

## PART 1

**Procedure:** Do the following:

1. Grind about 5 grams of peat in a mortar.
2. Pour the peat into a beaker. Put on safety goggles.
3. Heat the peat until it turns to ash. Keep the room well ventilated.
4. Set aside the beaker to cool.
5. Weigh the ash. How much ash is left in the beaker?

   ______________________

6. Describe the ash. ______________________

   ______________________

## PART 2

Repeat the procedure for lignite, bituminous and anthracite samples. This will take several class periods. Complete the following items:

### *Lignite*

1. Weigh the ash. How much ash is left in the beaker? ______________________
2. Describe the ash. ______________________

   ______________________

### *Bituminous*

1. Weigh the ash. How much ash is left in the beaker? ______________________
2. Describe the ash. ______________________

   ______________________

***Anthracite***

1. Weigh the ash. How much ash is left in the beaker? ____________________

2. Describe the ash. ____________________

____________________

## PART 3

Answer these items:

1. What advantage is there in burning hard coal rather than soft coal? ____________

2. What advantage is there in burning soft coal rather than hard coal? ____________

3. List three reasons some people choose not to burn coal.

   a. ____________________

   b. ____________________

   c. ____________________

4. What is coke? ____________________

   ____________________

5. If coal is used in place of petroleum, what will happen to our coal supply? ________

   ____________________

   What might be two problems future generations would face?

   a. ____________________

   b. ____________________

6. What three things should people know before they use fossil fuels for energy?

   a. ____________________

   b. ____________________

   c. ____________________

Name ______________________________ Date ______________________

# 2-10 Simulated Solar Cooker

**Introduction:** Sunlight is a renewable energy source. Conversely, energy from fossil fuels is nonrenewable. Once such fuels are used up, they cannot be replaced. In this activity you will use a 40-watt light source to represent the sun.

**Objective:** To build a model solar cooker.

**Materials:** Various sized beakers, funnel, aluminum foil, candle, microscope slides, glue, wooden splints, thread, mirror, tape, magnifier, white paper, colored construction paper, water, thermometer, and 40-watt lamp.

**Procedure:** Do the following:

1. Fill a beaker with 50 ml of water. Take the temperature of the water. The temperature is

   ______°C.
2. Now build a solar cooker using *only* the listed materials and your imagination.

   Complete these items on the back of this sheet.

1. Record the *highest* temperature reading. The temperature is ______°C.
2. When you finish building and testing the solar cooker, sketch your product. *Label* each part of the sketch.
3. Describe how you were able to raise the room temperature of the water. In other words, tell what you did to get the water to absorb more light energy from the heat source.
4. Why is a real solar cooker more efficient than the one you built?
5. In your opinion, what are three things people could do to conserve energy?
6. Several years ago a popular slogan was, "Don't be fuelish." What do you think the slogan meant?
7. Design a slogan or sign that you would like to see on car bumpers that would encourage people to conserve energy.

# Section 2: Energy Resources
# TEACHER'S GUIDE AND ANSWER KEY

## 2-1 Energy Resources Word Search

As an added activity, you can have students group the terms according to various categories—for example, oil/natural gas formation, coal formation, nuclear energy, and so on. Have students create their own labels for each group and use every term on the list. For example: *coal formation*—coal, bituminous, anthracite, energy, fossil fuels, lignite, organic, peat, and sedimentary.

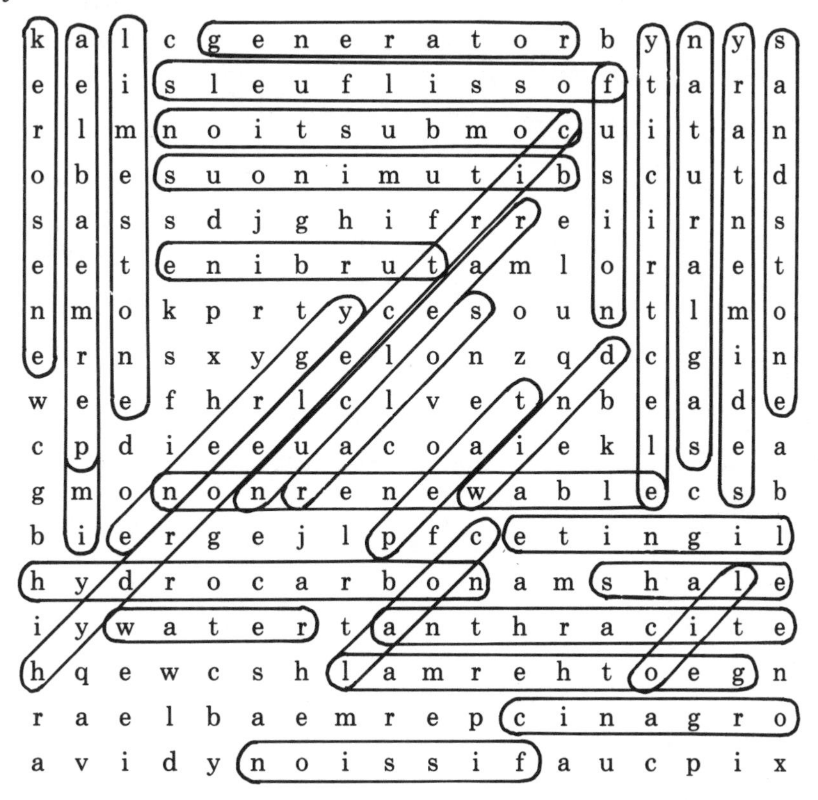


## 2-2 A Nonrenewable Energy Source: Coal

### *Part 1*

1. "black gold," h; 2. bituminous, j; 3. carbon, i; 4. chemical, f; 5. peat, b; 6. pressure, g; 7. swamp, k; 8. metamorphic, l; 9. sedimentary, a; 10. organic, c; 11. anthracite, e; 12. lignite, d

### *Part 2*

Order of sentences: 2, 4, 6, 1, 5, 3

## 2-3 Hydrocarbon Puzzle

1. anthracite; 2. hydrogen; 3. sedimentary; 4. carbon; 5. petroleum; 6. coal; 7. peat; 8. propane; 9. bituminous; 10. resources; 11. lignite; 12. shale

***Possible Answers for How Each Term Is Related to Hydrocarbons:***

1. Anthracite—composed of organic elements (H and O)
2. Hydrogen—an element that makes up hydrocarbons
3. Sedimentary—coal, a hydrocarbon product; a sedimentary rock
4. Carbon—an element that makes up hydrocarbon
5. Petroleum—a hydrocarbon product
6. Coal—a hydrocarbon product
7. Peat—the first stage of coal development; a hydrocarbon product
8. Propane—a petroleum product; a hydrocarbon
9. Bituminous—the third stage of coal development; a hydrocarbon product
10. Resources—coal, petroleum, and natural gas (all hydrocarbon products)
11. Lignite—the second stage of coal development; a hydrocarbon product
12. Shale—petroleum, a hydrocarbon product, can form in shale rock; shale becomes the source of oil.

## 2-4 Energy Source Puzzle

**Across:** 3. wind generator; 4. natural gas; 5. solar; 7. petroleum; 9. tidal power; 10. synfuels; **Down:** 1. hydroelectric; 2. coal; 6. nuclear; 8. geothermal

## 2-5 Energy Concern

Have a supply of newspapers and magazines in the classroom. Articles should deal with energy issues. Some students have trouble writing factual statements, so write two or three examples on the chalkboard to help get them started.

(Answers will vary.)

## 2-6 Fuel of the Future

### *Part 1*

1. d; 2. a; 3. b; 4. a; 5. d; 6. c; 7. b

### *Part 2*

1. geothermal; 2. fossil; 3. electricity; 4. biomass fuel; 5. synfuels

### *Part 3*

1. d; 2. e; 3. c; 4. f; 5. b; 6. a

## 2-7 Duster or Gusher?

The term *foraminifera* might confuse students. Tell them that foraminifera (forams, for short) are tiny, shell-forming protozoans that thrived in ocean waters during the Mississippian period. Geologists use these fossils as indicators for the presence of oil-bearing strata.

As an additional activity to enhance students' learning of the terms, make an overhead transparency of the gameboard and project it onto a screen. When a student lands on a particular symbol, he or she must define the term (or go back some spaces if defined incorrectly).

## 2-8 Making Hydrocarbon Models

Some students may wish to build three-dimensional hydrocarbon models. If so, provide them with styrofoam balls, black and yellow paint, brushes, and toothpicks. Some teachers prefer to use small marshmallows for carbon and hydrogen atoms. You may want to hand out ditto sheets with circles already drawn; this would save time.

## 2-9 Burning Coal

Keep the windows open and fresh air circulating throughout the room. This activity will take several class periods to complete.

### *Part 1*

Answers will vary.

### *Part 2*

Answers will vary.

### *Part 3*

1. Hard coal produces less ash than soft coal.
2. Soft coal tends to burn easier than hard coal.
3. a. It is dirty.
   b. It gives off a strong odor.
   c. Coal bins take up too much room.
4. It is the residue of coal after the gas has been expelled.
5. The coal supply will disappear earlier than predicted.
   a. The coal remaining would be very expensive.
   b. Atmospheric pollution might increase rapidly.
6. a. It is a nonrenewable energy source.
   b. It is a source of air pollution.
   c. Once it is gone, it may be very expensive to find and develop an alternate source of energy.

## 2-10 Simulated Solar Cooker

If certain items on the materials list are not available, replace them with other items. You could include aluminum soft drink cans, cardboard, wire, small pebbles, charcoal, and so on.

1. Answers will vary; 2. Sketches will vary; 3. Answers will vary; 4. Solar cookers are built by technicians who use expensive equipment; 5. a. Use only as much energy as needed, b. car pool, c. don't run air conditioners needlessly, d. don't leave lights burning when not in use, etc.; 6. Don't waste fuel, minimize car travel by forming car pools, don't take unnecessary car trips; 7. Slogans or signs will vary.

# Section 2: Energy Resources
## MINI-ACTIVITIES

Listed below are 15 mini-activities, one to five minutes in length, for students to do at the beginning or end of the period.

1. Unscramble the scrambled letters in the sentences below.

   a. When atoms are split during a chain reaction, aerulnc ________________ yegern ________________ is released.

   (**Answer:** nuclear energy)

   b. Sunlight produces rsaol ________________ yegner ________________.

   (**Answer:** solar energy)

   c. Compressed peat is called brown coal or eltiign ________________.

   (**Answer:** lignite)

2. Show, using arrows and circles, how fission occurs.

   (**Answer:** 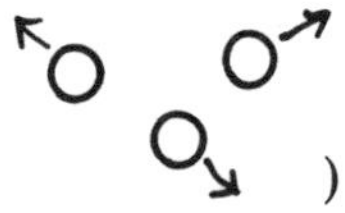 )

3. Show, using arrows and circles, how fusion occurs.

   (**Answer:** 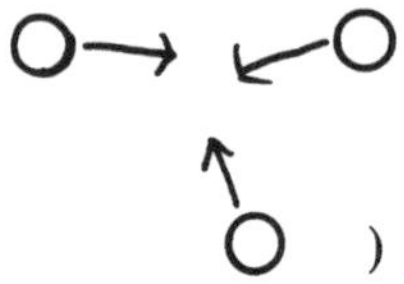 )

4. **Riddle:** Where does a nuclear scientist like to go on vacation?

   (**Answer:** fission)

5. Uncover the secret message:

   T + he +  + is + first letter + so +  + of + "in" + "eer" + gee + 4 + wind + and + hydropower.

   (**Answer:** The sun is a source of energy for wind and hydropower.)

6. **Riddle:** Miss C and Mr. H got married. They had six children: two girls and four boys. Show, using a chemical formula, what their family portrait might look like.

   (**Answer:** $C_2H_4$)

7. Name the nonrenewable energy resources that rhyme with the following words:

   a. luminous ________________
   d. mass ________________
   b. feet ________________
   e. stewed ________________
   c. linoleum ________________

   (**Answers:** a. bituminous; b. peat; c. petroleum; d. gas; e. crude)

8. **Riddle:** What does this drawing represent?

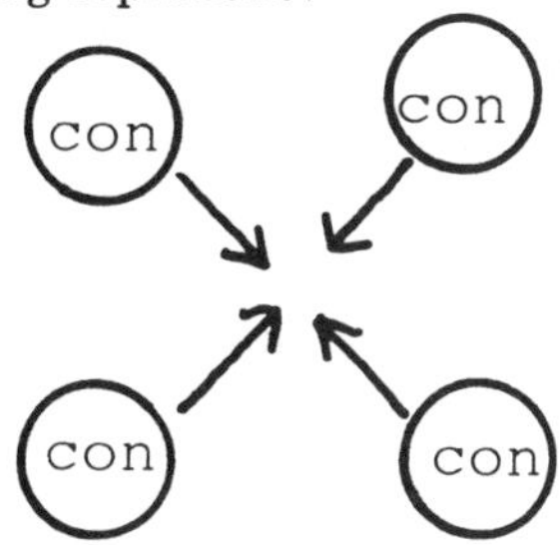


(**Answer:** confusion)

9. Fill in the spaces with renewable energy resources.

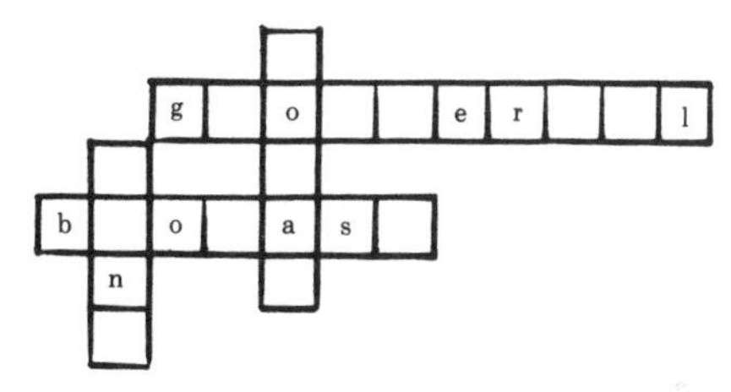

(**Answers:** *Down*—wind and solar; *Across*—geothermal and biomass)

10. Draw a straight line from the scrambled terms in Column A to the descriptions in Column B. (Answers are already shown.)

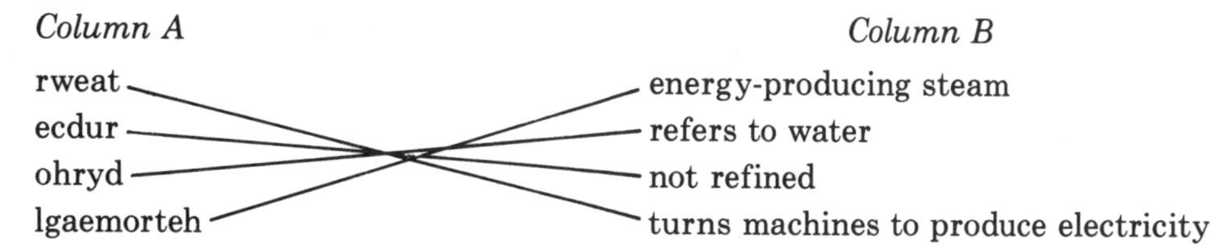


(Column A, top to bottom: water, crude, hydro, and geothermal)

11. List eight energy sources that have two connecting vowels in their names.

(**Answers:** biomass, bituminous, coal, fission, fuel, fusion, geothermal, nuclear, oil, petroleum, and so on)

12. **Riddle:** What would you have if you removed sol from solar?

(**Answer:** nothing)

13. **Problem:** Light travels 186,000 miles per second. The Earth is 93 million miles from the sun. If the sun suddenly burned out, how long would it be before we experienced darkness?

(**Answer:** 93 million divided by 186,000 equals 500 seconds. Five hundred divided by 60 (seconds) equals 8.3 minutes. The answer is 8.3 minutes.)

14. See how many first names or shortened forms of boys' names you can make from the word *geothermal.* You can use a letter more than once.

(**Answers:** George, Herm, Mel, Reg, Oral, Leo, Tom, Theo, Omar, Mat, Earl, Merle, Otto, Al, Matt, and so on)

15. Use your imagination and knowledge of science to identify the sketch below. *Hint:* It is made of silicon and produces current when sunlight strikes it.

(**Answer:** solar cell)

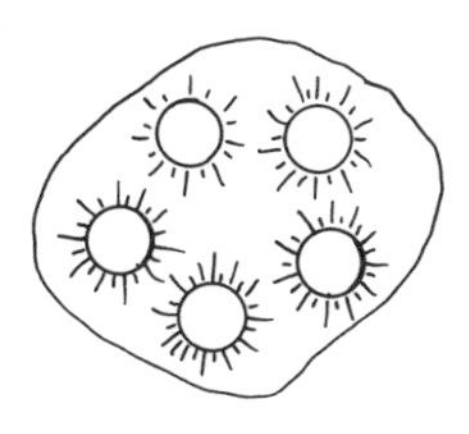

# Section 3
# MINERALS

## OUTLINE

3-1 Mineral Term Game
3-2 A Look at Minerals
3-3 Mineral Properties
3-4 Mineral Identification
3-5 Mineral Word Search
3-6 How Are Minerals Important to Us?
3-7 Hardness
3-8 Streak
3-9 Luster
3-10 Specific Gravity
3-11 Find the Mystery Mineral
3-12 Crystals
3-13 Microcrystals

## MATERIAL

The following laboratory materials are needed for Section 3:

- ammonium chloride solution
- balance
- beakers, 200 ml
- cold dilute hydrochloric acid
- copper coins
- crystals (halite, quartz, calcite)
- dropper bottles
- files
- fingernail clippings (supplied by students)
- glass plates (microscope slides)
- hammers
- lab aprons
- microscopes
- mineral samples (calcite, quartz, hematite, olivine, serpentine, pyroxene, pyrite, and feldspar, plus other samples as desired)
- paper towels
- pencil and paper
- potassium dichromate solution
- safety goggles
- salt solution
- spring balances
- steel knives
- streak plates
- string
- unglazed tile plates
- water

Name ______________________ Date ______________________

# 3-1 Mineral Term Game

Write the word that best fits the description on the left. When you are finished, the boxed letters will answer the question: What represents the flowers of the mineral kingdom? (a two-word answer).

1. Tendency of minerals to break in certain directions _ _ _ _ ☐ _ _ _
2. A mineral's resistance to being scratched _ ☐ _ _ _ _ _ _ _
3. Number 4 on the scale of hardness _ _ _ _ ☐ _ _ _
4. The ability of a mineral to hold together _ _ _ _ _ ☐ _ _
5. A colorless form of mica _ _ _ _ ☐ _ _ _ _
6. The way a mineral breaks _ _ _ _ _ ☐ _ _
7. A mineral that will bend and spring back _ _ _ ☐ _ _ _
8. Number 3 on the scale of hardness _ _ _ ☐ _ _ _
9. The color of a thin layer of finely powdered mineral _ _ ☐ _ _ _
10. Major ore of the metal beryllium _ _ _ ☐ _
11. Another name for olivine _ _ _ _ ☐ _ _ _ _ _
12. Also known as soapstone ☐ _ _ _
13. Orthoclase belongs in this group _ _ _ _ _ _ ☐ _
14. How a mineral shines ☐ _ _ _ _ _
15. A major element in quartz ☐ _ _ _ _ _

Name ______________________________ Date ______________________

# 3-2 A Look at Minerals

## PART 1

Complete the story by filling in the blanks with the correct term of the choices in parentheses.

What are minerals? They are ________ (symbols, elements) or compounds that occur ________ (naturally, spontaneously) in the Earth's ________ (mantle, crust) in ________ (solid, liquid), crystalline states.

Minerals can usually be identified by their physical properties. Some ________ (fracture, cleave) along smooth planes, like ________ (sulfur, mica) and ________ (calcite, hematite). Minerals vary in hardness, color, streak, and specific gravity.

Most minerals are crystalline. Their ________ (atoms, protons), ________ (electrons, ions), or ________ (neutrons, molecules) are arranged in ________ (fixed, unstable) patterns.

Specimens of the same mineral may vary greatly in ________ (tenacity, color) and ________ (physical, interior) appearance.

## PART 2

Unscramble the name of the mineral in the left-hand column and write the correct term in the space to the left of the number. Then draw a line connecting the name of the mineral to the mineral characteristics in the right-hand column.

| | | |
|---|---|---|
| ________ | 1. liathe | a. perfect cleavage in one direction; hardness 2.5 to 3 |
| ________ | 2. agenal | b. very heavy; metallic luster |
| ________ | 3. anircabn | c. cubic cleavage; ore of lead |
| ________ | 4. tiebtio | d. very soft, greasy, black streak |
| ________ | 5. apiregth | e. yields mercury; scarlet streak |
| ________ | 6. rutzaq | f. rock salt; cubic crystals |
| ________ | 7. lvirse | g. chemical formula $SiO_2$; hardness 7 |

Name ______________________ Date __________

# 3-3 Mineral Properties

Complete this crossword puzzle using words relating to mineral properties.

### Across

4. Uranium gives off radiations that can be detected with a Geiger counter
6. The breakage of a mineral so it yields definite flat surfaces
8. A mineral that will bend and spring back
9. A mineral that can be bent without breaking and will stay bent
10. A number indicating the ratio of the weight of a mineral to that of an equal volume of water (two words)
12. A mineral that can be drawn out into a wire
14. A mineral that can be flattened without breaking
15. A mineral that will fizz or bubble when touched with hydrochloric acid
17. The manner in which a mineral breaks
19. The way ordinary light reflects from a mineral's surface
20. The taste of halite

### Down

1. Mineral particles held loosely together
2. The color of a powdered mineral
3. Lodestone is said to be this
5. A mineral's resistance to scratching
7. The hue or appearance of a mineral
11. The emission of colored light rays by a substance during exposure to ultraviolet light
13. The continued giving off of light rays from certain minerals after exposure to ultraviolet light
16. A mineral that can be cut into shavings with a knife

Name ______________________ Date ______________

# 3-4 Mineral Identification

Complete the descriptions of the 15 minerals listed below. Each mineral has only one feature listed after it. The box contains identifiable characteristics needed to complete the descriptions. Choose the letters of two characteristics that fit each mineral and write them in the spaces to the right of the mineral. Some characteristics may be used more than once; others may not be used at all.

| | |
|---|---|
| a. specific gravity: 2.5 to 3.1 | k. has a soapy feel |
| b. hardness: 1.5 to 2.5 | l. hardness: about 2.5 |
| c. known as fool's gold | m. contains calcium |
| d. green carbonate of copper | n. specific gravity: 2.0 to 2.5 |
| e. specific gravity: 5.0 to 5.5 | o. formed by the weathering of feldspar |
| f. number 7 on scale of hardness | p. a tin ore |
| g. number 4 on scale of hardness | q. specific gravity: 2.9 to 3.2 |
| h. specific gravity: 8 or above | r. known as olivine |
| i. hardness: 5 to 6.5 | s. number 6 on scale of hardness |
| j. fizzes on contact with weak hydrochloric acid | t. specific gravity: about 5.0 |

1. Talc: __, __, hardness of 1
2. Calcite: __, double refraction, __
3. Quartz: Glassy or greasy luster, __, __
4. Gypsum: Number 2 on scale of hardness, __, __
5. Dolomite: __, __, hardness of 3.5 to 4
6. Kaolin: __, hardness between 1 and 2.5, __
7. Sulfur: Mostly yellow in color, __, __
8. Feldspar: __, vitreous to pearly luster, __
9. Pyrite: __, __, has a high metallic luster
10. Hornblende: A ferromagnesian silicate, __, __
11. Fluorite: __, __, octahedral cleavages
12. Magnetite: Black streak, __, __
13. Mica: __, __, pearly luster
14. Cinnabar: Ore of mercury, __, __
15. Hematite: __, __, cherry-red streak

Name ______________________________ Date ______________________

# 3-5 Mineral Word Search

Locate and circle the names of minerals commonly found in earth science textbooks. The names appear in the list below. Names may be found backward, forward, vertically, horizontally, and diagonally. Be careful!

```
E A C A L C I T E C F P N O A
S B H E T I T A M E H E I L T
U G A U O L E C I R T H Q E M
L T L A A S B E L I M A I U G
F V C P S E T I R A B W F T Q
U C O I J I K O E A N E L A G
R I P B M A U E U W A S C L E
M E Y O G L A X Z C G I E C T
E A R L F R I E Q N M B P M I
A H I Q M T A M C E T I N O L
C L T P E G D L O G C O E L A
O F E L D S P A R N Q T C B H
R X C U M U S P Y G I I P O T
Y Q U A R T Z I O R G T L S I
O S I L M E R E U P B E E T C
D T A M O N D Z A E R U B Y A
S L A O C W A E T I H P A R G
```

| | | | |
|---|---|---|---|
| azurite | chalcopyrite | graphite | mica |
| barite | chromite | gypsum | opal |
| bauxite | feldspar | halite | quartz |
| biotite | fluorite | hematite | sulfur |
| calcite | galena | limonite | talc |

Name ______________________ Date ______________________

# 3-6 How Are Minerals Important to Us?

There are about 2,500 known minerals. Many of these minerals provide the materials for industry. Some develop into beautiful gemstones; others provide valuable ingredients for building materials, cosmetics, household items, and so on. Minerals contribute to our comfort, happiness, and to the progress of the world.

Find out how much you know about minerals. After each mineral listed below, circle the lettered statement(s) that indicate(s) its importance.

1. Bauxite: a. lead ore; b. tin ore; c. aluminum ore
2. Halite: a. used in fertilizer; b. origin of salt; c. food product
3. Graphite: a. a fuel; b. source of pencil lead; c. a lubricant
4. Kernite: a. produces soap; b. used in welding; c. gemstone
5. Serpentine: a. decorative rock; b. insulation; c. fireproofing
6. Kaolinite: a. filler for paper; b. clay product; c. food source
7. Talc: a. used for paints; b. talcum powder; c. gun powder
8. Apatite: a. produces sulfur; b. fertilizer; c. nuclear reactor
9. Cuprite: a. iron ore; b. copper ore; c. silver ore
10. Cinnabar: a. produces mercury; b. gemstone; c. oil product
11. Tourmaline: a. gemstone; b. fossil fuel; c. insecticide
12. Muscovite: a. insulation; b. fireproofing; c. gemstone
13. Calcite: a. cement; b. food preservative; c. lead ore
14. Sulfur: a. sulfuric acid; b. fireworks; c. paper products
15. Topaz: a. gemstone; b. cosmetics; c. food processing
16. Quartz: a. pottery; b. semiprecious stones; c. glass
17. Gypsum: a. gemstone; b. plaster of Paris; c. food supplement
18. Cassiterite: a. uranium ore; b. vanadium ore; c. tin ore
19. Albite: a. pottery; b. writing utensils; c. gemstone
20. Carnotite: a. uranium ore; b. vanadium ore; c. paint

Name ______________________________ Date ______________________

# 3-7 Hardness

**Introduction:** Hardness is the resistance of a material to scratching. The scale of hardness ranges from 1 to 10, with 1 referring to the softest mineral and 10 the hardest mineral. The scale of hardness is as follows:

| | |
|---|---|
| 1. talc | 6. orthoclase |
| 2. gypsum | 7. quartz |
| 3. calcite | 8. topaz |
| 4. fluorite | 9. corundum |
| 5. apatite | 10. diamond |

These numbers merely indicate relative hardness. In other words, number 5 in the scale of hardness is not five times as hard as number 1, and so on.

**Objective:** To test the relative hardness of several minerals.

**Materials:** Mineral samples, fingernail, knife or steel file, copper coin, and small glass plate.

**Procedure:** Test each mineral sample using the following relative hardness testing scale:

1,2: Minerals scratched by the fingernail have a hardness of 2.5 or less.

3: Minerals can be cut easily by a knife; just scratched by a copper coin; not scratched by a fingernail.

4: Minerals can be scratched by a knife or file without difficulty but are not easily cut.

5: Minerals can be scratched by a knife or file with difficulty.

6: Minerals cannot be scratched by a knife; can be scratched by a file; will not scratch ordinary glass.

7: Minerals scratch glass easily.

8–10: Minerals seldom encountered for consideration.

List the names of the mineral samples and their relative hardness.

| | Mineral | Relative Hardness |
|---|---|---|
| 1. | ______________________ | ___ |
| 2. | ______________________ | ___ |
| 3. | ______________________ | ___ |
| 4. | ______________________ | ___ |
| 5. | ______________________ | ___ |
| 6. | ______________________ | ___ |
| 7. | ______________________ | ___ |
| 8. | ______________________ | ___ |

Name ______________________ Date ______________________

# 3-8 Streak

**Introduction:** Streak is the color of a mineral when powdered. This color is often different from the outside or external color. Soft minerals are tested by rubbing them on an unglazed tile plate. The undersides of flooring tiles make excellent streak plates. Hard minerals can be tested by scratching them with a file or by powdering them with a hammer.

**Objective:** To determine the streak of various minerals.

**Materials:** A variety of soft and hard minerals, unglazed tile plates, file, hammer, and safety goggles.

**Procedure:** Determine the streak of soft minerals by rubbing them against the rough surface of an unglazed tile plate. Record the external color and streak color in the spaces below. Determine the streak of hard minerals by scratching them with a file or powdering them by striking their surfaces with a hammer. (Wear goggles to protect your eyes from flying debris.) Record the external color and streak color in the spaces below.

| | Mineral | External Color | Streak Color |
|---|---|---|---|
| 1. | | | |
| 2. | | | |
| 3. | | | |
| 4. | | | |
| 5. | | | |
| 6. | | | |
| 7. | | | |
| 8. | | | |

## Questions

1. Which minerals had the *same* external color and streak?
2. Which minerals had a streak *different* from their external colors?
3. After completing the activity, what can you say about a mineral's streak and external color?

Name ______________________ Date ______________________

# 3-9 Luster

**Introduction:** Mineralogists refer to the surface appearance of a mineral in reflected light as *luster.* Therefore, how a mineral shines in reflected light is its luster.

Metals such as gold, silver, copper, aluminum, and lead are said to have a metallic luster or shine. Minerals that exhibit a nonmetallic luster vary considerably. The terms used to describe the various kinds of nonmetallic luster originate from objects such as glass, resin, wax, grease, pearls, satin, and silk. The chart below lists luster or shine and the product from which the description originated.

| Luster or Shine | Product |
|---|---|
| Metallic | Silver, copper, gold, etc. |
| Adamantine | Oily surface (flashy, diamond-like, or greasy) |
| Vitreous | Broken glass |
| Resinous | Rosin or amber |
| Waxy | Pieces of wax |
| Pearly | Imitation pearls or Mother of Pearl |
| Satiny | Silk or satin |
| Dull or earthy | Soil |

**Objective:** To describe the luster or shine of various minerals.

**Materials:** Various metallic and nonmetallic minerals.

**Procedure:** Examine each mineral and complete the following chart.

| | Mineral | Luster or Shine |
|---|---|---|
| 1. | ______________________ | ______________________ |
| 2. | ______________________ | ______________________ |
| 3. | ______________________ | ______________________ |
| 4. | ______________________ | ______________________ |
| 5. | ______________________ | ______________________ |
| 6. | ______________________ | ______________________ |
| 7. | ______________________ | ______________________ |
| 8. | ______________________ | ______________________ |

Name ______________________________ Date ____________________

# 3-10 Specific Gravity

**Introduction:** The specific gravity or density of a mineral is a comparison of its weight with that of water—a way of determining how many times as heavy as water a mineral is. Quartz represents an average specific gravity with a 2.65 reading. In comparison, gypsum (2.2 to 2.4) would be light; olivine (3.2 to 3.6) heavy; pyrite (5.0) very heavy. The formula for specific gravity is

$$\text{Specific Gravity} = \frac{\text{Dry Mineral Weight}}{\text{Dry Mineral Weight} - \text{Wet Mineral Weight}}$$

**Objective:** To determine the specific gravity of several minerals.

**Materials:** Balance, minerals (light, heavy, very heavy), beaker (200 ml), string, water.

**Procedure:** Determine the specific gravity of each mineral in the following manner:

1. Weigh each mineral on an accurate balance.
2. Suspend the mineral from the balance by a piece of string and weigh again while the mineral is completely submerged in a beaker of water.
3. Avoid resting the mineral against the side of the beaker. Subtract the weight in water from the dry weight. The difference is the weight of a volume of water equal to the volume of the mineral.
4. Divide the weight of an equal volume of water into the dry weight (weight in air). The final product is the specific gravity.

Record your measurements in the following chart.

| | Mineral | Dry Weight | Difference in Dry/Wet Weight | Divide Difference into Dry Weight | Specific Gravity |
|---|---|---|---|---|---|
| 1. | | | | | |
| 2. | | | | | |
| 3. | | | | | |
| 4. | | | | | |
| 5. | | | | | |
| 6. | | | | | |
| 7. | | | | | |
| 8. | | | | | |

**Questions**

1. How many minerals were light? ____________
2. How many minerals were heavy? ____________
3. How many minerals were very heavy? ____________

Name ______________________ Date ______________________

# 3-11 Find the Mystery Mineral

**Introduction:** To identify a specific mineral, certain clues must be given and the proper tests completed. Few minerals can be identified by color, luster, hardness, or specific gravity alone. Therefore, several tests are necessary to pinpoint or isolate a selected mineral.

**Objective:** To use clues and perform tests necessary to identify a mystery mineral correctly.

**Materials:** Eight minerals—calcite, quartz, hematite, feldspar, olivine, serpentine, pyrite, pyroxene—spring balance, beaker (200 ml), string, water, glass plates (microscope slides), steel file, knife, copper coin, fingernail, streak plate, lab aprons, and cold dilute hydrochloric acid.

**Procedure:** Use the clues to test all eight minerals. If the minerals are tested accurately, one mineral—the mystery mineral—will be revealed.

Test the mineral samples using the following clues, and record your answers in the spaces below.

***Clue 1:***

The mystery mineral's specific gravity is between 2.5 and 2.7.

Possible mystery minerals are ______________, ______________, ______________,

______________, and ______________. (You may have less than five possible answers.)

***Clue 2:***

The mystery mineral's hardness ranges from 2.5 to 4.0.

Possible mystery minerals are ______________, ______________, and ______________.

***Clue 3:***

The mystery mineral's streak is white.

Possible mystery minerals are ______________, ______________, and ______________.

Now ask your teacher for the cold dilute hydrochloric acid. If you can't decide which of the remaining specimens is the mystery mineral, put a drop of acid on each specimen. The mystery mineral will effervesce (fizz) readily in cold dilute hydrochloric acid.

**The mystery mineral is ______________.**

Name ______________________ Date ______________________

# 3-12 Crystals

**Introduction:** Crystals are naturally formed particles with regular shapes, flat surfaces, and straight edges. A mineral's internal structure, which consists of atoms, ions, and molecules, arranges itself in definite patterns. Calcite, halite, galena, pyrite, quartz, and mica are examples of crystals which display various crystal shapes.

**Objective:** To examine how calcite, halite, and quartz crystals are alike and how they are different.

**Materials:** Quartz, calcite, and halite crystals, pencil and paper, glass plates (microscope slides), cold dilute hydrochloric acid, lab aprons, dropper bottle, streak plate, spring balance, water, and string.

**Procedure:** Perform the necessary experiments to answer the following questions. Write your answers on another sheet of paper.

1. All crystals—quartz, calcite, and halite—have a vitreous or glassy luster. If you submerge each crystal in water, which crystal will show a different luster? How has the luster changed?
2. How do calcite and quartz compare in taste?
3. How is the taste of halite different from the taste of calcite and quartz?
4. Lift each crystal one at a time. List which crystal feels the heaviest, next heaviest, and lightest. Now do the specific gravity test on each crystal. Describe how your first list compared with the specific gravity test results.
5. Briefly describe the shape of each crystal. How are they alike? How are they different?
6. Draw two dots close together on a piece of paper. Place each crystal on top of the dots. Do the dots change position? If so, explain what you think caused the phenomenon.
7. Place a drop of cold dilute hydrochloric acid on each crystal. Which crystal reacts to the acid? Explain what reaction takes place. Why do you think this reaction took place? Rinse acid off of crystals and wipe dry.
8. Does calcite scratch quartz, or does quartz scratch calcite?
9. Does halite scratch calcite, or does calcite scratch halite?
10. Which crystal(s) will scratch glass?
11. Rank the crystals according to their hardness—the first mineral on the list is softest, and so on.
12. Can you tell these crystals apart by their streak alone? Why or why not?
13. List five things all of these crystals have in common.
14. List three ways in which these crystals are different.

Name ________________________________ Date ________________________

# 3-13 Microcrystals

**Introduction:** An interesting way to study crystals is to watch them grow under a microscope. In this activity, you will place a prepared chemical solution on a microscope slide, heat the bottom of the slide, and see crystals come out of the solution. Write and draw your answers on another sheet of paper.

**Objective:** To observe crystals growing under a microscope.

**Materials:** Ammonium chloride solution, salt solution, potassium dichromate solution, microscope, microscope slides, dropper bottles, goggles, paper towels, lab aprons, matches, and water.

## PART 1

Place two or three drops of ammonium chloride solution on the center of a slide. Gently warm the bottom of the slide with a match until the solution begins to crystallize around the edges. Place the slide under a microscope, using low power.

1. Sketch the crystals.
2. Describe how the crystals appear to grow.
3. List three characteristics of ammonium chloride crystals.

## PART 2

Place two or three drops of salt solution on the center of a slide. Gently warm the bottom of the slide with a match until the solution begins to crystallize around the edges. Place the slide under a microscope, using low power.

1. Sketch the crystals.
2. Describe how the crystals appear to grow.
3. List three characteristics of salt solution crystals.

## PART 3

Place two or three drops of potassium dichromate solution on the center of a slide. Gently warm the bottom of the slide with a match until the solution begins to crystallize around the edges. Place the slide under a microscope, using low power.

1. Sketch the crystals.
2. Describe how the crystals appear to grow.
3. List three characteristics of potassium dichromate crystals.

## PART 4

Examine each of the crystal slides under low power when the solution dries completely.

1. How are all the crystals alike?
2. How do the crystals differ?

When you finish the activity, wash and dry the slides before putting them away.

# Section 3: Minerals
# TEACHER'S GUIDE AND ANSWER KEY

## 3-1 Mineral Term Game

1. cleavage
2. hardness
3. fluorite
4. tenacity
5. muscovite
6. fracture
7. elastic
8. calcite
9. streak
10. beryl
11. chrysolite
12. talc
13. feldspar
14. luster
15. silica

Boxed letters form: VARIOUS CRYSTALS.

## 3-2 A Look at Minerals

### *Part 1*

elements; naturally; crust; solid; cleave; mica; calcite; atoms; ions; molecules; fixed; color; physical

### *Part 2*

1. halite, f; 2. galena, c; 3. cinnabar, e; 4. biotite, a; 5. graphite, d; 6. quartz, g; 7. silver, b

## 3-3 Mineral Properties

**Across:** 4. radioactive; 6. cleavage; 8. elastic; 9. flexible; 10. specific gravity; 12. ductile; 14. malleable; 15. effervesce; 17. fracture; 19. luster; 20. salty. **Down:** 1. brittle; 2. streak; 3. magnetic; 5. hardness; 7. color; 11. fluorescence; 13. phosphorescence; 16. sectile; 18. transparent

## 3-4 Mineral Identification

1.a, k; 2. j, m; 3. f, a; 4. n, m; 5. m, a; 6. o, a; 7. b, n; 8. a, s; 9. t, c; 10. i, q; 11. g, m; 12. i, e; 13. 1, a; 14. b, h; 15. i, e

## 3-5 Mineral Word Search

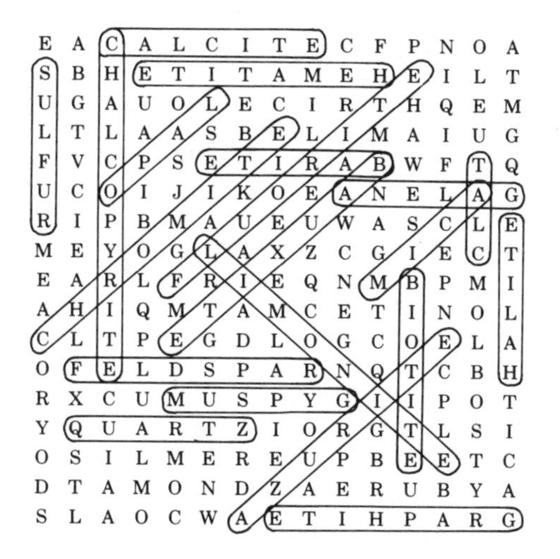

## 3-6 How Are Minerals Important to Us?

1. c; 2. b, c; 3. b, c; 4. a, b; 5. a, c; 6. a, b; 7. a, b; 8. b; 9. b; 10. a; 11. a; 12. a, b; 13. a; 14. a, b, c; 15. a; 16. a, b, c; 17. b; 18. c; 19. a; 20. a, b

## 3-7 Hardness

The mineral hardness will vary with each individual sample. Try to find a mixture of soft and hard minerals. If you want students to experiment with more than eight minerals, have them record their answers on the back of the worksheet.

## 3-8 Streak

Inexpensive tile plates can be purchased at most paint and art supply stores. If you want students to experiment with more than eight minerals, have them record their answers on the back of the worksheet.

1. Answers will vary; 2. Answers will vary; 3. They are not always the same.

## 3-9 Luster

Luster can present a problem for students. Some students are not familiar with terms like resinous, vitreous, and adamantine. If possible, show everyday items that demonstrate these luster features. If you want students to experiment with more than eight minerals, have them record their answers on the back of the worksheet.

## 3-10 Specific Gravity

The specific gravity of a sample mineral varies according to its density. For example, a light mineral will show a lower reading than a more dense mineral. Students should take their time, make accurate measurements, and be able to perform basic math. Some students will need extra help in dividing decimals. If you want students to experiment with more than eight minerals, have them record their answers on the back of the worksheet.

*Hint:* Have the students take *gram* measurements, not ounces. Their answers will be more accurate.

## 3-11 Find the Mystery Mineral

Avoid substituting minerals unless you test them first to see if they match the clues.

Clue 1: calcite, feldspar, quartz, serpentine; Clue 2: serpentine, calcite; Clue 3: calcite, serpentine; Mystery mineral: calcite

*Note:* You can prepare a dilute hydrochloric acid solution by adding 15 ml of concentrated hydrochloric acid to 100 ml of water. Have students perform each test in the order given in the activity.

## 3-12 Crystals

1. Halite; luster turned from vitreous to dull.
2. Both are tasteless.
3. Halite has a salty taste.
4. Answers will vary.
5. Halite—cube-like; quartz—six-sided; calcite—rhombohedral. They are all solid, transparent crystals. They differ in shape, specific gravity, hardness, and streak.

6. Yes. Calcite shows four dots instead of two. Calcite crystals of a transparent variety give a double image when light passes through them. This is known as double refraction.
7. Calcite. It reacts by fizzing or effervescing. The calcium in calcite reacts with the hydrochloric acid.
8. Quartz scratches calcite.
9. They both have approximately the same hardness. They may scratch each other. Chances are, however, that calcite will scratch halite.
10. Quartz scratches glass.
11. Softest—halite or calcite; hardest—quartz
12. No. They all show a colorless or white streak.
13. They are all solid, transparent substances. They have the same streak and luster. They are all useful to people.
14. They have different chemical compositions, specific gravities, and different shapes.

## 3-13 Microcrystals

You can easily prepare each solution by covering the bottom of a dropper bottle with the chemical, filling the bottle three-fourths full of water, and shaking it vigorously. If your solution is too dilute, add more chemical. This worksheet needs more than one class period to complete.

### *Part 1*

1. Sketches will vary; 2. Fast. They cover the slide with a snowflake pattern; 3. Solid, has a definite shape, white color.

### *Part 2*

1. Sketches will vary; 2. Fast. They clump together; 3. Solid, has a cube-like shape, white color.

### *Part 3*

1. Sketches will vary; 2. Fast. They clump together. Some form long, branch-like patterns; 3. Solid, rectangular shape, yellowish-orange color.

## Section 3: Minerals
## MINI-ACTIVITIES

Listed below are eight mini-activities, one to five minutes in length, for students to do at the beginning or end of the period.

1. Unscramble the mineral names and tell what they all have in common.

   IAECCTL ZRAQTU ILTHEA

   (**Answer:** calcite, quartz, and halite. They are all minerals in solid form. They are crystals, and some share common characteristics.)

2. Fill in the blanks with the correct letters. Put the circled letters together to complete the statement below.

   1. Science of minerals _ i n ◯ r ◯ l _ g _
   2. Crystals breaking in a definite direction Ⓒ _ e _ _ a _ e
   3. Earthy and dull are examples of this property ◯ _ s ⓣ e _
   4. Salt crystals have this shape ◯ u _ ⓘ c a _

   Clear _ _ _ _ _ _ _ is called Iceland Spar.

   (**Answers:** 1. mineralogy; 2. cleavage; 3. luster; 4. cubical. Answer to the statement is calcite.)

3. Complete the crossword puzzle.

   ### Across

   2. Fool's gold
   3. Number 7 on the scale of hardness
   5. An example of tenacity

   ### Down

   1. Black mica
   4. Number 6 on the scale of hardness
   5. Another way to describe fracture

   (**Answers:** *Across*—2. pyrite; 3. quartz; 5. brittle; *Down*—1. biotite; 4. feldspar; 5. break)

4. Complete the mineral names by filling in the missing letters. Then unscramble the letters to answer the question: What is the term for no light being transmitted through a crystal?

   1. _uartz
   2. To_az
   3. Flu_rit_
   4. _ranium
   5. Hem_tite

   (**Answer:** qpoeua = opaque)

5. Two minerals have the same size, shape, hardness, luster, and specific gravity. Yet both can be easily separated. How would you explain this?

   (**Answer:** They have different colors.)

6. Think of five minerals that have the combination "al" in their names.

   (**Answers:** Talc, aluminum, opal, calcite, realgar, copalite, chalcopyrite, malachite, galena, halite, etc.)

7. Mary tested the specific gravity of two different pieces of gypsum. One piece read 2.52; the other piece measured 2.68. Assuming Mary did the tests correctly, why did she receive different readings?

   (**Answer:** One sample may have been purer than the other.)

8. Think of five girls' names that come from gemstones.

   (**Answers:** Pearl, Amber, Ruby, Sapphire, Garnet, Opal, etc.)

# Section 4
# ROCKS

## OUTLINE


4-1 Sedimentary Rock Puzzle
4-2 Metamorphic Rock Scramble
4-3 Igneous Rock Maze
4-4 Rock Word Search
4-5 Matching Rock Properties
4-6 Rhyming Rock Puzzle
4-7 Sedimentary Rock Observation
4-8 Igneous Rock Observation
4-9 Metamorphic Rock Observation
4-10 Determining the Density of a Rock


## MATERIAL

The following laboratory materials are needed for Section 4:

- balances
- beakers, 250 ml
- graduated cylinders, 100 ml
- hand lens or magnifiers
- igneous rock samples (granite, basalt, pumice, obsidian, scoria)
- metamorphic rock samples (slate, marble, serpentine, quartzite)
- overflow cans
- pencils or pens
- scraping tools
- sedimentary rock samples (limestone, sandstone, shale, conglomerate)
- thread
- water

Name ______________________________ Date ______________________

# 4-1 Sedimentary Rock Puzzle

Complete the puzzle using the names of sedimentary rocks. First fill in the blanks in the words below the puzzle, and you'll have the answers to the puzzle.

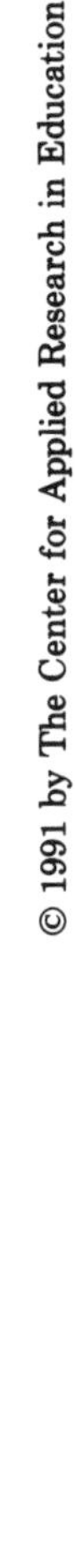


**Across**

1. s _ _ _ s _ _ n e
4. _ _ a l _
5. _ a r _
8. r _ _ k _ a _ t (two words)
9. o _ l _ t e
11. _ i _ e _ t _ n _
12. c _ n _ l o _ _ r _ _ _

**Down**

2. _ _ a v e r _ _ n e
3. c h _ _ _
6. _ i l _ s _ o n e
7. _ o l _ m _ _ e
10. _ r e c _ _ a

sedimentary

Name ______________________________ Date ____________________

# 4-2 Metamorphic Rock Scramble

## PART 1

Unscramble the letters to find the name of the metamorphic rock or the term related to the metamorphic process. Write each word, one letter in each blank. On the back of this sheet, write a description of each word.

1. eissgn

2. utiqetazr

3. ldfeoita

4. hssstci

5. emlarb

6. sholernf

_ _ _ _ _ _ _ _ ◯

metamorphic

7. doefntolain

8. srtzyrcdeelila

9. ltase

10. epesrurs

## PART 2

Unscramble the circled letters to complete the following statement:

Metamorphic changes occur while the rock is in a solid or _ _ _ _ _ _ _ _

_ _ _ _ _. (two words)

Name ______________________________ Date ______________________

# 4-3 Igneous Rock Maze

Several igneous rock names appear in the puzzle. The same name may appear more than once. Names may appear vertically, horizontally, backward, and forward. To solve the maze, you must draw a line connecting the complete rock names *or part* of each name, and you may use the same name more than once. Begin the line at the START position and end at the EXIT position. If you run into a dead end, retrace your steps.

START

| t | a | r | p | u | m | i | c | e | t | l | a | s | a | b |
|---|---|---|---|---|---|---|---|---|---|---|---|---|---|---|
| o | b | s | i | d | i | a | n | t | u | t | n | c | n | z |
| d | i | o | r | i | t | e | a | i | f | a | d | o | d | t |
| a | n | d | e | s | i | t | e | r | f | l | e | r | e | a |
| e | i | f | g | a | b | b | r | o | o | c | s | i | s | e |
| t | s | a | n | d | r | w | l | i | b | p | i | a | i | p |
| i | a | i | c | c | e | r | b | d | s | d | t | v | t | r |
| n | p | u | m | i | c | e | k | q | i | i | e | t | e | h |
| a | t | u | f | f | c | m | r | l | d | o | p | l | w | y |
| r | h | y | o | l | i | t | e | n | i | r | s | i | a | o |
| g | c | o | a | l | a | y | t | e | a | i | x | s | t | l |
| a | i | c | c | e | r | b | i | a | n | i | o | a | e | i |
| b | h | j | a | n | d | e | s | i | t | e | f | f | u | t |
| b | s | c | o | r | i | a | l | g | u | c | l | a | y | e |
| r | h | y | o | l | i | t | e | z | f | t | r | e | h | c |
| o | e | c | i | m | u | p | f | a | f | e | t | l | a | s |

EXIT

List in alphabetical order the igneous rock names that you connected. You need not repeat a name.

______________________ ______________________

______________________ ______________________

______________________ ______________________

igneous

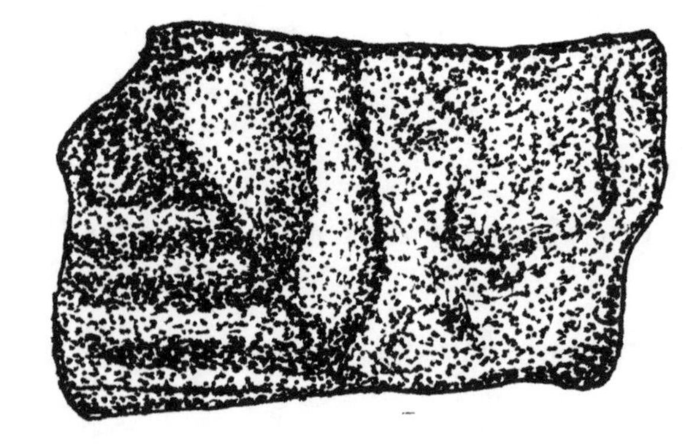

Name ______________________ Date ______________

# 4-4 Rock Word Search

Locate and circle the 18 rocks commonly found in earth science textbooks. The rock names are listed below the puzzle. They may be found backward, forward, vertically, horizontally, and diagonally.

T B A S A L T A C G A B B R O B E
S F I G H E T I T O D I R E P J L
I K C L O P S U X A R N Q S M Q I
H E C I M U P A S E T V H E U C M
C I E S Z A W E N I Y A E A H E E
S A R S G K J I A D L C R E A T S
L O B I A E T O F E S T R M I I T
R S S I E N G A F I Z T E R A L O
M A Y T E C D F U I A M O T L O N
E Z I P S W P E T O L I V N S Y E
M A R B L E Z E S E D T Q A E H A
A E I O V T S S L E F N R O H R I
S A P N X E T A R E M O L G N O C

| | | |
|---|---|---|
| basalt | gneiss | quartzite |
| breccia | hornfels | rhyolite |
| chert | limestone | schist |
| conglomerate | marble | serpentine |
| diorite | peridotite | shale |
| gabbro | pumice | tuff |

Name ____________________ Date ____________________

# 4-5 Matching Rock Properties

Rocks are usually divided, according to their origin, into three groups: (1) igneous, (2) sedimentary, and (3) metamorphic. Igneous rocks come from a molten state, sedimentary rocks are composed of previously existing rocks, and metamorphic rocks are the result of heat, pressure, and chemical action on igneous and sedimentary rocks.

On the right is a list of rock properties. On the left is a list of rocks. Match the rocks with their properties by writing the appropriate letters from the right-hand column next to each rock. In addition, write *I* for *Igneous, S* for *Sedimentary,* and *M* for *Metamorphic* in the box.

__ 1. granite ☐

__ 2. rhyolite ☐

__ 3. limestone ☐

__ 4. gabbro ☐

__ 5. scoria ☐

__ 6. conglomerate ☐

__ 7. slate ☐

__ 8. basalt ☐

__ 9. marble ☐

__ 10. diorite ☐

__ 11. shale ☐

__ 12. quartzite ☐

a. intrusive; coarse-grained; composed mostly of plagioclase and pyroxene

b. a coarse, frothy rock filled with spherical holes

c. foliated; microscopic grains; originates from shale and siltstone

d. extrusive; fine-grained; contains mica, olivine, pyroxene, and plagioclase

e. granular; composed of plagioclase and small amounts of iron minerals

f. made of mud, clay, and silt; fragments less than 1/16 mm

g. formed from quartz sandstone

h. cemented quartz and rock fragments

i. fine-grained; mixture of feldspar and quartz

j. nonclastic; composed of calcite or microscopic shells

k. nonfoliated; origin: pure limestone

l. light-colored; granular texture; mostly feldspar and quartz

Name ______________________________ Date ______________________

# 4-6 Rhyming Rock Puzzle

## PART 1

Use the two rhyming words to help you find the correct rock. Write the name of the rock in the space provided. Then complete the statement that follows.

1. alert, flirt, ______________
2. ______________, assault, halt
3. ______________, mist, fist
4. ice, ______________, mice
5. hail, ______________, rail
6. ______________, rhinestone, grindstone
7. cellulite, ______________, dynamite
8. rate, date, ______________
9. meridian, ______________, median
10. ______________, gargle, warble
11. planet, ______________, Janet
12. insight, delight, ______________
13. ______________, escrow, skidrow
14. dumb-bells, bluebells, ______________
15. guillotine, ______________, Byzantine
16. ______________, rough, enough

There are __________ igneous rocks, __________ sedimentary rocks, and __________ metamorphic rocks listed above.

## PART 2

The missing terms below relate to the study of rocks. Use the descriptive statements and rhyming words in parentheses to find the answers. This is a tough one!

1. The instrument used to study rock features is known as a ______________ microscope. (spectrographic, demographic)
2. To identify an unknown rock, a student must be thoroughly familiar with __________-______________ minerals. (sock-swarming, flock-storming)
3. Donna picked up a particularly attractive rock and showed it to her friend, Shirley. "Shirley," she said, "this rock is covered with inorganic bodies having definite geometrical forms." "Wow," responded Shirley, "I'll bet that you become a ______________." (geographer, oceanographer)

Name ______________________ Date ______________________

# 4-7 Sedimentary Rock Observation

**Introduction:** Sedimentary rocks are an accumulation of rock particles that settle into horizontal layers and slowly unite together into rocks. These rock fragments may be deposited by wind, water, ice, or other means.

**Objective:** To examine the physical features of sedimentary rocks.

**Materials:** Chart, pencil or pen, hand lens or magnifier, scraping tool, and sedimentary rock samples (limestone, sandstone, shale, and conglomerate).

## PART 1

**Procedure:** Do the following:

1. Carefully examine each rock sample, one at a time.
2. Fill in the Sedimentary Rock Observation Chart. Use the terms *clastic, organic,* and *crystalline* to describe texture. Here is a brief description of these terms:

   Clastic: Composed of particles and fragments cemented together

   Organic: Composed of an accumulation of shells, plant remains, bones, and so forth, cemented together

   Crystalline: Composed of crystals crowded together

In addition, use the terms *light, medium,* and *heavy* to designate relative weight.

## PART 2

Answer these questions:

1. Why do you think some sandstones are harder than others?

   ______________________

2. What do you think determines the texture of conglomerate?

   ______________________

3. Name three sedimentary rocks used for decorative landscape or building materials.

   ____________, ____________, ____________

## Sedimentary Rock Observation Chart

| Name | Color | Weight | Texture | Other Features | Sketch |
|---|---|---|---|---|---|
| | | | | | |
| | | | | | |
| | | | | | |
| | | | | | |
| | | | | | |

Name ______________________________ Date ______________________

# 4-8 Igneous Rock Observation

**Introduction:** Igneous rocks are molten rocks that cooled and solidified. The molten rock, magma, may reach the surface of the Earth through lava flows or ejected fragments thrown out of an exploding volcano. Some magma never reaches the surface. When magma cools, it turns into rock known as igneous rock.

**Objective:** To examine the physical features of igneous rock.

**Materials:** Chart, pencil or pen, hand lens or magnifier, scraping tool, and igneous rock samples (granite, basalt, pumice, obsidian, and scoria).

## PART 1

**Procedure:** Do the following:

1. Carefully examine each rock sample, one at a time.
2. Fill in the Igneous Rock Observation Chart. Use the terms *pyroclastic, glassy, aphanitic,* and *granular* to describe texture. Here is a brief description of these terms:

   Pyroclastic: Made up of volcanic glass, pumice, broken fragments, and crystals cemented together

   Glassy: Composed mostly of massive or streaky volcanic glass

   Aphanitic: Composed of tiny crystals which give the rock an earthy or dull luster

   Granular: Composed of crystals large enough to see without a lens or magnifier

In addition, use the terms *light, medium,* and *heavy* to designate relative weight.



## PART 2

Answer these questions:

1. Why are fossils not found in igneous rocks? ______________________________

   ______________________________________________________________

2. What determines the texture of igneous rocks? ______________________________

   ______________________________________________________________

3. Name three igneous rocks used for decorative landscape or building materials.

   ____________________, ____________________, ____________________

## Igneous Rock Observation Chart

| Name | Color | Weight | Texture | Other Features | Sketch |
|---|---|---|---|---|---|
| | | | | | |
| | | | | | |
| | | | | | |
| | | | | | |
| | | | | | |

Name ______________________________ Date ____________________

# 4-9 Metamorphic Rock Observation

**Introduction:** Metamorphic rocks are rocks whose original form has been changed by heat, pressure, or chemical action. When a rock goes through metamorphic change (metamorphism), its texture, mineral composition, or both change.

**Objective:** To examine the physical features of metamorphic rocks.

**Materials:** Chart, pencil or pen, hand lens or magnifier, scraping tool, and metamorphic rock samples (slate, marble, serpentine, and quartzite).

## PART 1

**Procedure:** Do the following:

1. Carefully examine each rock sample, one at a time.
2. Fill in the Metamorphic Rock Observation Chart. Use the terms *foliated* and *nonfoliated* to describe texture. Here is a brief description of these terms:

   Foliated: Mineral crystals or grains form in parallel layers or bands. The rock splits readily or peels along these layers or bands.

   Nonfoliated: Rocks without layers or bands

In addition, use the terms *light, medium,* and *heavy* to designate relative weight.

## PART 2

Answer these questions:

1. Why are some metamorphic rocks hard to identify? ______________________________

   ______________________________________________________________

2. Why do foliated rocks break more easily than nonfoliated rocks?

   ______________________________________________________________

   ______________________________________________________________

3. Name three metamorphic rocks used for decorative landscapes or building materials.

   ____________________, ____________________, ____________________

## Metamorphic Rock Observation Chart

| Name | Color | Weight | Texture | Other Features | Sketch |
|---|---|---|---|---|---|
| | | | | | |
| | | | | | |
| | | | | | |
| | | | | | |

Name ______________________________ Date ____________________

# 4-10 Determining the Density of a Rock

**Introduction:** Density is the mass of a body divided by its volume. How heavy a rock feels is determined by its density. For example, a chunk of granite has a much greater density than a piece of pumice of the same volume. Density is written as a ratio, such as grams per cubic centimeter ($g/cm^3$).

**Objective:** To compare the densities of different rocks.

**Materials:** Chart, pencil or pen, overflow can, thread, graduated cylinder (100 ml), balance, beaker (250 ml), water, and assorted rocks (granite, basalt, sandstone, shale, marble, quartzite, and so on).

## PART 1

**Procedure:** Do the following:

Determine the density of each rock. Place all measurements on the Rock Density Chart. Here are the steps for determining density:

- Weigh the rock on the balance (g).
- Fill the overflow can to the top of the rim. Try to keep the water from spilling over the rim.
- Place a beaker directly under the spout of the can.
- Tie a string around the rock. Slowly submerge the rock until it is completely covered with water. Collect the water that overflows in the beaker.
- Pour the overflow water from the beaker into a graduated cylinder. The volume of water collected in the cylinder will be the volume of the rock ($cm^3$).
- Divide the volume ($cm^3$) into the mass (weight) of the rock (g). The answer will be the density of the rock ($D = M/V$).

*Example:* Rock *A* weighs 42 grams. Its volume is 16 ml. Forty-two grams divided by 16 ml equals 2.625 or 2.63. Therefore, Rock *A* has a density of 2.63.

## PART 2

Answer these questions on the back of this sheet.

1. What is meant by a low-density rock?
2. What is meant by a high-density rock?
3. How is mass different from weight?
4. The three most common rocks are shale, granite, and basalt. According to your measurements (if you used these examples in the activity), which rock has the highest-density reading?
5. Name three conditions that could cause your measurements to be inaccurate.
6. What do you think could cause low-density magma to become high-density lava?

## Rock Density Chart

| Name | Mass Weight (g) | Volume ($cm^3$) | Mass Divided by Volume | Density ($g/cm^3$) |
|---|---|---|---|---|
| | | | | |
| | | | | |
| | | | | |
| | | | | |
| | | | | |
| | | | | |
| | | | | |

# Section 4: Rocks
# TEACHER'S GUIDE AND ANSWER KEY

## 4-1 Sedimentary Rock Puzzle

*Across:* 1. sandstone; 4. shale; 5. marl; 8. rock salt; 9. oolite; 11. limestone; 12. conglomerate; *Down:* 2. travertine; 3. chert; 6. siltstone; 7. dolomite; 10. breccia

## 4-2 Metamorphic Rock Scramble

### *Part 1*

1. gneiss; 2. quartzite; 3. foliated; 4. schists; 5. marble; 6. hornfels; 7. nonfoliated; 8. recrystallized; 9. slate; 10. pressure

### *Descriptions*

1. a granite-like rock with layers; 2. a rock that comes from quartz sandstone; 3. rocks that show banding, such as gneiss and slate; 4. rocks that show some recrystallization; 5. metamorphosed limestone; 6. a rock composed of shales; 7. rocks without banding, such as marble and quartzite; 8. rocks that have undergone repeated recrystallization and do not show foliation; 9. metamorphosed shale; 10. a constraining, depressing force partly responsible for twisting and buckling rock

### *Part 2*

plastic state

## 4-3 Igneous Rock Maze

These terms appear in the puzzle: basalt, tuff, granite, diorite, felsite, obsidian, pumice, breccia, rhyolite, gabbro, andesite, and scoria.

A possible solution to the puzzle:

START

| t | a | r | p | u | m | i | c | e | t | l | a | s | a | b |
|---|---|---|---|---|---|---|---|---|---|---|---|---|---|---|
| o | b | s | i | d | i | a | n | t | u | t | n | c | n | z |
| d | i | o | r | i | t | e | a | i | f | a | d | o | d | t |
| a | n | d | e | s | i | t | e | r | f | l | e | r | e | a |
| e | i | f | g | a | b | b | r | o | o | c | s | i | s | e |
| t | s | a | n | d | r | w | l | i | b | p | i | a | i | p |
| i | a | i | c | c | e | r | b | d | s | d | t | v | t | r |
| n | p | u | m | i | c | e | k | q | i | i | e | t | e | h |
| a | t | u | f | f | c | m | r | l | d | o | p | l | w | y |
| r | h | y | o | l | i | t | e | n | i | r | s | i | a | o |
| g | c | o | a | l | a | y | t | e | a | i | x | s | t | l |
| a | i | c | c | e | r | b | i | a | n | i | o | a | e | i |
| b | h | j | a | n | d | e | s | i | t | e | f | f | u | t |
| b | s | c | o | r | i | a | l | g | u | c | l | a | y | e |
| r | h | y | o | l | i | t | e | z | f | t | r | e | h | c |
| o | e | c | i | m | u | p | f | a | f | e | t | l | a | s |

EXIT

The rocks in alphabetical order are: breccia, diorite, felsite, gabbro, pumice, and rhyolite.

## 4-4 Rock Word Search

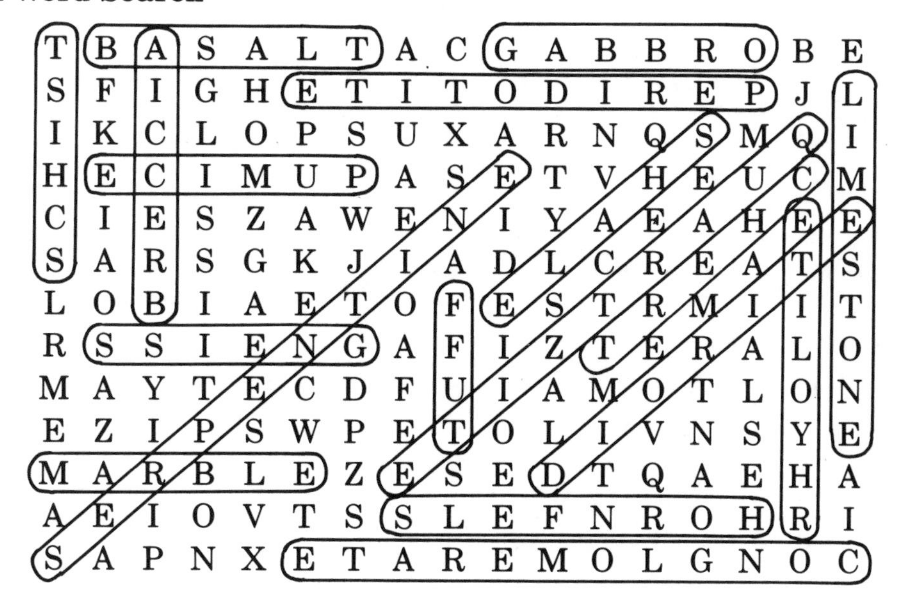


## 4-5 Matching Rock Properties

1. l (I); 2. i (I); 3. j (S); 4. a (I); 5. b (I); 6. h (S); 7. c (M); 8. d (I); 9. k (M); 10. e (I); 11. f (S); 12. g (M)

## 4-6 Rhyming Rock Puzzle

### *Part 1*

1. chert; 2. basalt; 3. schist; 4. gneiss; 5. shale; 6. limestone; 7. dolomite; 8. slate; 9. obsidian; 10. marble; 11. granite; 12. phyllite; 13. gabbro; 14. hornfels; 15. serpentine; 16. tuff

There are five igneous rocks (basalt, obsidian, tuff, gabbro, granite), four sedimentary rocks (shale, chert, limestone, dolomite), and seven metamorphic rocks (schist, gneiss, slate, marble, phyllite, hornfells, serpentine).

### *Part 2*

1. petrographic; 2. rock-forming; 3. crystallographer

## 4-7 Sedimentary Rock Observation

The rocks mentioned are representative samples. Feel free to substitute other rocks if needed. Do the same for activities 4-8 and 4-9.

### *Part 1*

Observation answers will vary. Answers for the texture column are: sandstone—clastic; limestone—clastic or organic; shale—clastic; conglomerate—clastic.

### *Part 2*

1. The harder sandstone grains are held together with stronger cement; i.e., silica rather than iron oxide or calcium carbonate.

2. The size of the fragments in it
3. Limestone, sandstone, conglomerate, shale, etc.

## 4-8 Igneous Rock Observation

### *Part 1*

Observation answers will vary. Answers for the texture column are: granite—granular; basalt—aphanitic; pumice—glassy; obsidian—glassy; scoria—aphanitic.

### *Part 2*

1. They would be destroyed in a molten mass.
2. The size or shape of crystals in the rock
3. Obsidian, pumice, scoria, basalt, granite, etc.

## 4-9 Metamorphic Rock Observation

### *Part 1*

Observation answers will vary. Answers for the texture column are: slate—foliated; quartzite—nonfoliated; marble—nonfoliated; serpentine—nonfoliated.

### *Part 2*

1. They have been recrystallized to the point at which the nature of the original rock is impossible to determine.
2. They are held loosely together along parallel lines, thus allowing for easy breakage.
3. Serpentine, slate, marble, etc.

## 4-10 Determining the Density of a Rock

You can make overflow containers out of soda cans or milk cartons. Try to use a mixture of sedimentary, igneous, and metamorphic rocks in this lab. Again, the rocks mentioned are the easiest specimens to obtain, but don't limit the students to these. Be ready to assist students who have trouble reading a balance or dividing decimals.

### *Part 1*

Measurements will vary.

### *Part 2*

1. A rock composed of loosely compacted minerals, such as pumice
2. A rock composed of tightly compacted minerals, such as basalt
3. Mass is the measurement of the amount of material in an object. Weight is the gravitational force that the Earth exerts on an object.
4. Answers will vary; probably basalt.
5. Working too fast, misreading the balance or graduated cylinder, mathematical error, defective equipment, etc.
6. Low-density magma melts large amounts of high-density minerals as it flows upward. As it spews from the volcano, it cools and hardens into high-density lava.

## Section 4: Rocks
## MINI-ACTIVITIES

Listed below are 14 mini-activities, one to five minutes in length, for students to do at the beginning or end of the period.

1. You have five minutes to make a list of eight terms related to rocks from the letters listed in the box. You may use letters more than once.

| p | n | i | t | o | f | l | a | d | e | c | u | b | h | m | s | y | r |
|---|---|---|---|---|---|---|---|---|---|---|---|---|---|---|---|---|---|

   (**Possible Answers:** sedimentary, metamorphic, foliated, nonfoliated, schist, slate, shale, tuff, obsidian, pumice, limestone, sandstone, etc.)

2. What does it mean when someone says to another person, "Hey, do you know you're a real rock?"

   (**Answer:** a hardhead; dense upstairs)

3. Fill in the blanks with words that sound like the words in parentheses.

   "Sandra, it's ________________ for me to say this, but . . . um . . . I think you're very ________________." (tuff, gneiss)

4. **Riddle:** What igneous rock is found with ice in it?

   (**Answer:** pumice)

5. **Riddle:** What type of rock rhymes with mechanic?

   (**Answer:** volcanic)

6. **Riddle:** Which is the brightest igneous rock?

   (**Answer:** rhyolite)

7. What would you get if you chipped the end off a cinder block?

   (**Answer:** a chip off the old block)

8. What metamorphic rock resembles a reptile more than any other rock?

   (**Answer:** serpentine)

9. What would you have if you made a circle with 10 rocks?

   (**Answer:** a rock band)

10. **Riddle:** Betty wanted to make a cake out of earth materials. She used three main ingredients: pyroxene, amphibole, and serpentine. What kind of cake did Betty make?

    (**Answer:** marble)

11. **Riddle:** What rock has salt in it but doesn't taste salty?

    (**Answer:** basalt)

12. Complete the puzzle. All answers are fine-grained metamorphic rocks.

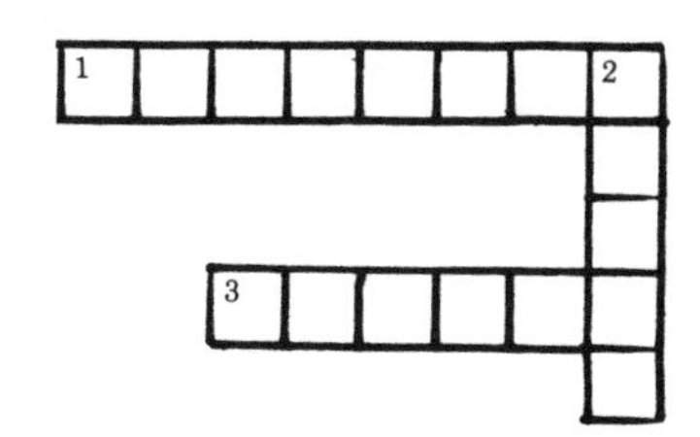


**Across**

1. Begins with the letter "H" (hornfells)
3. Foliated and recrystallized structure (schist)

**Down**

2. Made up of mica and quartz (slate)

13. Complete the puzzle. All answers are metamorphic rocks containing the letters "A" and "E."

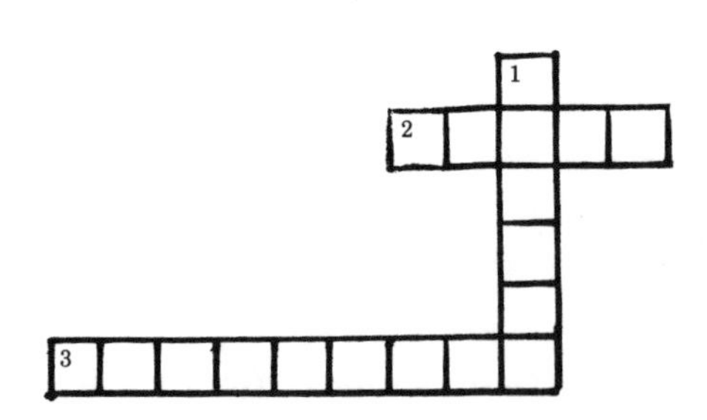


**Across**

2. Comes from shale (slate)
3. Comes from sandstone (quartzite)

**Down**

1. Comes from limestone (marble)

14. Complete the following rock descriptions. (Answers to be filled in are in parentheses.)

| Original Rock, Igneous or Sedimentary | Resulting Metamorphic Rock |
|---|---|
| granite | (gneiss) |
| (felsite or tuff) | schist |
| shale | (slate) |
| (basalt) | hornblende schist, talc schist, serpentine |
| sandstone | (quartzite) |

Section 5

# VOLCANOES

## OUTLINE

5-1 Volcano Vocabulary Puzzle
5-2 Fill-in Mystery Message
5-3 Anatomy of a Volcano
5-4 The L-term Volcanic Puzzle
5-5 Volcanic Rock Observation
5-6 Specific Gravity of Volcanic Rocks
5-7 Build a Model Volcano
5-8 Simulated Volcanic Rock Petroglyphs

## MATERIAL

The following laboratory materials are needed for Section 5:

- beaker tongs
- beakers (50, 100, 200 ml)
- bicarbonate of soda
- burners
- charcoal powder
- crayons
- Florence flask (250 ml)
- jar lids
- large paperclips
- magnifiers or hand lens
- masking tape
- metal probes (dissecting needles or nails)
- mixing jars (8 to 10 oz.)
- modeling clay
- mortars and pestles
- nails
- newspaper
- paper towels
- pencils or pens
- plaster of Paris
- red food coloring
- rulers
- sharp-pointed rocks
- spring balances
- stirrers
- string
- vinegar
- volcanic rock samples (pumice, obsidian, scoria, basalt, tuff, felsite [rhyolite], gabbro, granite)
- water
- writing paper

Name ______________________ Date ______________________

# 5-1 Volcano Vocabulary Puzzle

## PART 1

Each clue below refers to one of the words listed in Part 2. Select the word that best fits each clue, write it below, and circle it in the word search puzzle. *Hint:* The term *caldera* may be used twice.

1. Molten rock beneath the Earth's crust ______________________
2. Consolidated volcanic ash ______________________
3. A giant summit depression on a volcano ______________________
4. Molten material ejected from a volcano ______________________
5. A crack in the Earth's interior through which lava flows ______________________
6. A series or network of fissures ______________________
7. A vertical layer of hardened lava ______________________
8. A smooth, ropy Hawaiian lava ______________________
9. A rough, chunky Hawaiian lava ______________________
10. A sunken, collapsed crater at the top of a volcano ______________________
11. Lava materials deposited at the base of a volcano ______________________
12. A fan-shaped bank of deposited volcanic material ______________________
13. A gradual wearing away of rock materials ______________________
14. Rock materials greater than 1/16 mm in diameter ______________________
15. Smallest rock particles ______________________

16. Rock particles smaller than sand but larger than clay ______________________

17. Blobs of lava hurled from a volcano ______________________

18. When a volcano spouts out lava ______________________

19. A light, porous lava; may float on water ______________________

20. Mounds of viscous lava built over volcanic vents ______________________

21. Ancient carvings in volcanic rocks ______________________

22. Ancient pictures drawn with dye on volcanic rock ______________________

23. A horizontal layer of hardened lava ______________________

24. The main lava chamber in the neck of a volcano ______________________

25. A mountain formed by lava and rock material ejected from within the Earth's crust ______________________

26. The materials belched from the throat of a volcano ______________________

27. A steep-sided volcano formed from cinders and ash ______________________

28. A broad-domed volcano formed by lava flows from a vent or from fissures ______________________

29. A volcano that is intermediate in form between a cinder and shield volcano ______________________

30. A general term for granite rock formed at great depths ______________________

31. A general term for volcanic lavas; obsidian and pumice are examples ______________________

## PART 2

Use the terms listed below to match the clues given in Part 1. As noted in Part 1, one term—*caldera*—is used twice. The terms in the word search puzzle can be found forward, backward, horizontally, vertically, and diagonally.

| | | | |
|---|---|---|---|
| AA | ejecta | magma | sill |
| alluvial fan | erosion | Pahoehoe | silt |
| alluvium | erupts | petroglyphs | tuff cones |
| caldera | extrusive | pictographs | vent |
| cinder cone | fissure | pumice | volcanic domes |
| clay | intrusive | rift zone | volcano |
| composite cone | lava | sand | |
| dike | lava bombs | shield | |

a f a e p e v i s u r t x e k d e x p
b i s n a z p c m h e r u s s i f e a
s e n o c e t i s o p m o c t y t v e
c j a z l e o h e o h a p l a r a i p
m e p t a u a s u n s e r n o l o s q
h c u f y b g e i b v s o g s a y u l
l t m i u a j n m v t i l h o d a r l
k a i r m i e o i p s y i x n t e t i
o j c t u d b c u o p e f a n a c n s
n i e o i a t r r h l w s e m e z i o
a h s k v b e e s d e a v b a g b s p
c n e a u t a d i s g a r e d l a c m
l q l r l s e n o c f f u t r i w m o
o p a l l u v i a l f a n n w e c e c
v o l c a n i c d o m e s i l t l o s

Name ______________________________ Date ____________________

# 5-2 Fill-in Mystery Message

Fill in the missing letters. Then place the numbered letters in the correct blanks of the message to find the secret message.

V___(39)lcan___(41)c m___(5)un___(30)ai___(42)s may be big, up to 14,000 feet high. They are built from molt___(9)n lava ejec___(36)ed ___(32)rom the E___(34)rth's crust.

Volcanoes eject molt___(26)n lava, as___(7), gas, and s___(31)lid materia___(29)s. There are three basic ___(22)ypes of vol___(18)anoe___(2): cinder cone___(35), shield v___(17)lcano___(24)s, and c___(38)mpos___(20)te cones.

So___(3)e e___(10)amples of lava ___(25)oc___(19) are ___(4)b___(27)idian, ba___(21)alt, p___(28)mice, and granite.

Hawaiian volcanoes are mostly of the s___(23)ie___(40)d varie___(8)y. Shiel___(15) volcano___(14)s are made up of an accumul___(1)tion of very mobile lava. E___(16)___(12)pting a___(6) high temperature, the lava runs swi___(33)tly from the vent, sp___(13)eading widely, addin___(43) to a ___(37)one of gen___(11)le slope.

**Secret Message:** ___ ___ ___ ___ ___ ___ ___ ___ ___ ___ ___ ___ ___ ___ ___
1 2 3 4 5 6 7 8 9 10 11 12 13 14 15

___ ___ ___ ___ ___ ___ ___ ___ ___ ___ ___ ___ ___ ___ ___
16 17 18 19 20 21 22 23 24 25 26 27 28 29 30

___ ___ ___ ___ ___ ___ ___ ___ ___ ___ ___ ___ ___
31 32 33 34 35 36 37 38 39 40 41 42 43

Name ______________________ Date ______________________

# 5-3 Anatomy of a Volcano

Be constructive. On a separate piece of paper (8½″ by 11″), sketch a volcano using the recipe below. When you complete the diagram, title the volcano Mount (your name). Use any available earth science text to help you locate and label the listed structures on your sketch. You'll need a metric ruler, regular pencil, and a red or orange pen or pencil.

Here is a list of structures with recommended dimensions:

| Structure | Dimensions (centimeters) |
|---|---|
| Gas cloud | Length/width = flexible |
| Inactive vent | Length = 15 cm; width = 1 cm |
| Crater | Width = 1 cm |
| Active vent | Length = 15 cm; width = 1 cm |
| Lava flow | Length = 7–10 cm; width = ½ cm |
| Magma storage chamber (located at base of vent in crust | Length/width = flexible |
| Source of magma (connected to magma storage chamber located in mantle) | Length/width = flexible |

Begin by drawing an east/west line across the bottom of the page. Start the line about 5 centimeters from the bottom. Make the line at least 20 centimeters long. This line will represent the border between the crust and mantle.

Use the following color code on your sketch:

| | |
|---|---|
| Gas cloud | = pencil |
| Inactive vent | = pencil |
| Crater | = pencil |
| Active vent | = red or orange |
| Lava flow | = red or orange |
| Magma storage chamber | = red or orange |
| Source of magma | = red or orange |

Be creative. On the back of this sheet, write a make-believe description for each of the following situations.

1. How did the volcano behave the day before the eruption?
2. Describe how the volcano reacted during the eruption.
3. What did the area surrounding the volcano look like after the eruption?

Name ______________________________ Date ______________________

## 5-4 The L-term Volcanic Puzzle

Find and circle the 10 terms in the puzzle related to volcanoes. Each term contains the letter "L." In the spaces below the puzzle, list the terms in alphabetical order. Use the hints in parentheses to help you. The words can be found forward, backward, and diagonally.

1. ______________________________ (base, deposits)
2. ______________________________ (most common lava)
3. ______________________________ (sunken crater top)
4. ______________________________ (smallest particles)
5. ______________________________ (obsidian, for example)
6. ______________________________ (melted magma)
7. ______________________________ (green, glassy silicate)
8. ______________________________ (broad-domed volcano)
9. ______________________________ (hardened horizontal lava)
10. ______________________________ (lava mountain)

Name ______________________ Date ______________________

# 5-5 Volcanic Rock Observation

**Introduction:** Volcanic rocks differ according to their mineral content. Whether or not they contain quartz, feldspar, olivine, hornblende, biotite, pyroxene, or other mineral products depends on where the volcano erupts. A volcano's lava may be coarse or fine grained, porphyritic (a texture of fairly large crystals set in a mass of very fine crystals), or glassy—depending on the cooling rate. A rough-textured rock generally results from slow cooling. Conversely, a smooth-textured rock is the result of fast cooling.

**Objective:** To examine the various colors, weights, and textures of volcanic rock.

**Materials:** Magnifier or hand lens, pencil and paper, metal probe (dissecting needle or nail), balance, volcanic rock samples (pumice, obsidian, scoria, basalt, tuff, felsite, gabbro, granite, rhyolite, and so forth).

**Procedure:** Do the following:

Examine each rock specimen, one at a time. Fill in the chart below with your observations. Use the terms *light, medium,* and *heavy* to describe weight; *coarse* or *fine grained* to identify texture; *slow* or *fast* to describe cooling rate.

## Volcanic Rock Observation Chart

| Rock Name | Texture | Color | Cooling Rate | Weight |
|---|---|---|---|---|
| | | | | |
| | | | | |
| | | | | |
| | | | | |
| | | | | |
| | | | | |
| | | | | |
| | | | | |
| | | | | |
| | | | | |
| | | | | |
| | | | | |
| | | | | |

Name ______________________________ Date ____________________

# 5-6 Specific Gravity of Volcanic Rocks

**Introduction:** Igneous rocks form from magmas. When basaltic magma cools quickly, it forms a fine-textured, dark-colored rock known as basalt. When basaltic magma cools more slowly, it may form a coarse-textured, dark rock called gabbro.

If a granitic magma cools quickly, it may produce a fine-textured, light-colored rock known as rhyolite. If granitic magma cools more slowly, it may produce a dark-colored, coarse-textured rock known as granite.

In this activity you will determine the specific gravity or "heft" of various volcanic rocks. Specific gravity is the ratio of the weight of an object to the weight of an equal volume of water. Do all volcanic rocks have the same specific gravity? Let's find out.

## PART 1

**Objective:** To determine the specific gravity of various volcanic rocks.

**Materials:** Chart, beaker (100–200 ml), string, spring balance, pencil, volcanic rocks (gabbro, rhyolite, granite, basalt, scoria, tuff, and so forth).

**Procedure:** Do the following for each rock. Then record measurements on the recording chart.

1. Tie one end of string around the rock. Make a loop at the opposite end. Place the loop over the hook on the spring balance.
2. Weigh the rock sample suspended in air. Record the weight on the chart.
3. Immerse the rock in a beaker of water. Don't allow the rock to rest against the bottom of the beaker or along the sides.
4. Record the water weight on the chart. *Note:* The rock weighs less in water because of water's buoyant or lifting effect.
5. Substitute the measurements in the formula below:

$$\text{Specific Gravity} = \frac{\text{Weight of Mineral in Air}}{\text{Weight of Mineral in Air} - \text{Weight of Mineral in Water}}$$

For example:

$$\frac{35\text{ g}}{\underset{(15\text{ g})}{35\text{ g} - 20\text{ g}}}$$

35 Divided by 15 = 2.33
Specific Gravity = 2.33

## Volcanic Rock Recording Chart

| Rock | Air Weight (g) | Water Weight (g) | Air Weight – Wet Weight (Difference) | Air Weight Divided by Difference | Specific Gravity |
|---|---|---|---|---|---|
| | | | | | |
| | | | | | |
| | | | | | |
| | | | | | |
| | | | | | |
| | | | | | |
| | | | | | |

### PART 2

Answer these questions:

Which rock, basaltic or granitic, shows a higher specific gravity reading?

Why is this so?

Name ______________________________ Date ____________________

# 5-7 Build a Model Volcano

**Introduction:** Model volcanoes are fun to build and display. Once your creative juices begin to flow, you can build numerous models from materials found around your home. This activity describes two different volcano models, one that "erupts" and one of the extinct variety.

**Objective:** To build a model volcano.

**Materials:** Florence flask (250 ml), bicarbonate of soda, newspaper, charcoal powder, paper towels, vinegar, pencil or pen, modeling clay, crayons, pumice, mortar and pestle, burner, beaker (50 ml), and beaker tongs.

## PART 1

**Procedure:** Construct a "soda cone" in the following manner:

1. Fill a 250-ml Florence flask one-third full with bicarbonate of soda. Add enough charcoal powder to turn the soda black.
2. Cover the lab table or desk with paper towels.
3. Add enough vinegar to cause the reaction to flow out of the flask, down the sides, and onto the paper.
4. After the reaction stops, trace the flow pattern on the paper towels. Is there anything unique or strange about the pattern of flow? Why or why not?

   ______________________________________________________________

   ______________________________________________________________

5. Will the reaction occur again without adding more vinegar? Stir the mixture and find out.

## PART 2

Answer these questions. Use the back of this sheet if you need more space.

1. Heat and pressure within the Earth's crust force magma to reach the surface. What represents heat and pressure in the "soda cone"?

   ______________________________________________________________

2. Why might shaking the flask cause the mixture to "erupt" again?

   ______________________________________________________________

   ______________________________________________________________

3. What might cause a volcano to erupt several times in a short period of time?

   ______________________________________________________________

   ______________________________________________________________

## PART 3

Construct a clay model of a volcano in the following manner:

1. Spread newspaper over the work area.
2. Build a volcanic mountain using modeling clay. Make the model about 6 inches high and 3 to 4 inches wide at the base.
3. Mold a rugged terrain surrounding the mountain.
4. Slowly melt several dark-colored crayons (wrappers removed) in a 50-ml beaker over a burner. Pour the melted crayon over the model. The crayon simulates the flow of lava. *Note:* The melted crayon cools and hardens quickly.
5. Grind some pumice in a mortar and pestle. Pour the powder over the model. Push small chunks of pumice into the sides of the model. This will give the model a more "volcanic" look.
6. Present your model to the class and explain what it shows.



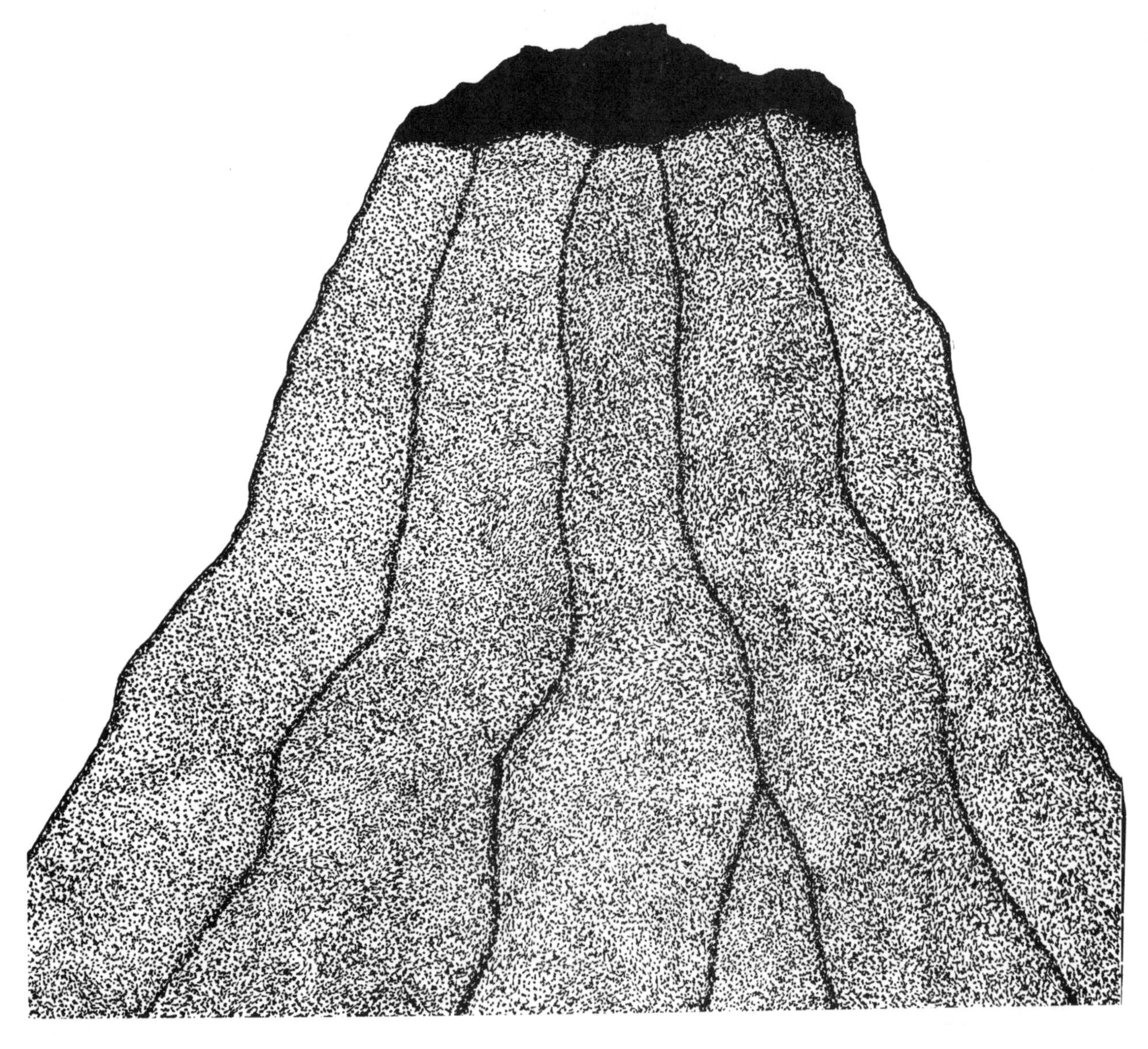

Name ______________________ Date ______________________

# 5-8 Simulated Volcanic Rock Petroglyphs

**Introduction:** Five miles south of Lahaina, Maui (in the Hawaiian Islands) lie the remnants of a volcanic lava flow that came to rest millions of years ago. These remnants provided early Polynesians with unique scribbling tablets. These people relied on volcanic boulders for tablets and lava chips or beach rocks for scrapers.

These stone carvings of pictures and symbols are known as petroglyphs. Petroglyphs are traces of a culture long gone, a prehistoric art thought by some to be anywhere from 200 to 2,000 years old, or older. The Hawaiians called them *Kaha Ki'i,* which means "drawn picture." Many Hawaiian petroglyphs depict tools, weapons, vessels, animals, and people. Scientists believe these carvings are attempts of earlier civilizations to communicate with one another and to tell their stories.

**Objective:** To reproduce simple figures and interpret their possible meanings. *Note:* Plaster of Paris will be used to simulate volcanic rock.

**Materials:** Two jar lids, plaster of Paris, mixing jar (8–10 oz.), stirrer, water, newspaper, paper and pencil, large paperclips, nails or rocks, red food coloring, ruler, and masking tape.

## PART 1

**Procedure:** Do the following:

1. Spread newspaper over the desk or lab table.
2. Place masking tape on the bottom of two jar lids. Write your name and class period on the tape.
3. Fill a jar half-full of plaster of Paris. Slowly add water to the plaster while stirring. Add two or three drops of red food coloring to darken the mixture. The red dye will simulate the color of lava rock. Stir until the mixture reaches milkshake consistency.
4. Pour the mixture into the jar lids. Gently tap the bottom of the lids against the desk or table while pouring the mixture. This helps the mixture spread out evenly and eliminates air pockets and bubbles.
5. Set jar lids aside to dry.

## PART 2

Read the following paragraph. Then sketch on paper the figures and symbols you feel best sum up the story. *Note:* Use the examples of petroglyph drawings provided here to help you get started. When the plaster casts dry, place them over newspaper and scratch the figures and symbols with a paperclip, nail, or sharp-pointed rock.

> *Pretend you live in the year 3500. Your civilization is crumbling, and you want to leave a record behind for others to find. You wish to tell a story, to show future generations how people lived in the year 3500. Unfortunately, for the last three centuries life has returned to a primitive state. All you have for leaving a record are plaster blocks and a scratcher.*

After you finish scratching, write a brief summary explaining what each figure and symbol means and why you chose these illustrations to tell your story.

Find a partner. Switch plaster casts. While you examine your partner's record, he or she studies your finished product. Do the following:

1. Place a piece of paper lengthwise on your desk. Using a ruler, draw a line from top to bottom in the middle of the paper. Write the word SCRATCHING in the left-hand column and the word MEANING in the right-hand column.
2. Duplicate each scratch mark you see on the plaster by drawing it in the left-hand column. Write what you think the scratch mark means in the right-hand column.
3. On the back of the page, write a brief (100 words or less) history of your partner's civilization based on the plaster recordings. Remember, interpret the historical events based on how you perceive the meaning of each figure.
4. When you finish writing the history, meet with your partner and share with each other what you discovered about each other's civilization.



**Examples of Petroglyph Drawings**

## Section 5: Volcanoes
## TEACHER'S GUIDE AND ANSWER KEY

### 5-1 Volcano Vocabulary Puzzle

#### *Part 1*

1. magma; 2. tuff cones; 3. caldera; 4. lava; 5. fissure; 6. rift zone; 7. dike; 8. Pahoehoe; 9. AA; 10. caldera; 11. alluvium; 12. alluvial fan; 13. erosion; 14. sand; 15. clay; 16. silt; 17. lava bombs; 18. erupts; 19. pumice; 20. volcanic domes; 21. petroglyphs; 22. pictographs; 23. sill; 24. vent; 25. volcano; 26. ejecta; 27. cinder cone; 28. shield; 29. composite cone; 30. intrusive; 31. extrusive

#### *Part 2*

a f a e p e v i s u r t x e k d e x p

b i s n a z p c m h e r u s s i f e a

s e n o c e t i s o p m o c t y t v e

c j a z l e o h e o h a p l a r a i p

m e p t a u a s u n s e r n o l o s q

h c u f y b g e i b v s o g s a y u l

l t m i u a j n m v t i l h o d a r l

k a i r m i e o i p s y i x n t e t i

o j c t u d b c u o p e f a n a c n s

n i e o i a t r r h l w s e m e z i o

a h s k v b e e s d e a v b a g b s p

c n e a u t a d i s g a r e d l a c m

l q l r l s e n o c f f u t r i w m o

o p a l l u v i a l f a n n w e c e c

v o l c a n i c d o m e s i l t l o s

### 5-2 Fill-in Mystery Message

Volcanic mountains may be big, up to 14,000 feet high. They are built from molten lava ejected from the Earth's crust.

Volcanoes eject molten lava, ash, gas, and solid materials. There are three basic types of volcanoes: cinder cones, shield volcanoes, and composite cones.

Some examples of lava rock are obsidian, basalt, pumice, and granite.

Hawaiian volcanoes are mostly of the shield variety. Shield volcanoes are made up of an accumulation of very mobile lava. Erupting at high temperature, the lava runs swiftly from the vent, spreading widely, adding to a cone of gentle slope.

*Message:* A smooth-textured rock is the result of fast cooling.

## 5-3 Anatomy of a Volcano

The list of structures are only recommendations. Allow students the freedom to experiment and try different dimensions, colors, and so on.

The make-believe descriptions will vary with each student.

## 5-4 The L-term Volcanic Puzzle

1. alluvium; 2. basalt; 3. caldera; 4. clay; 5. lava; 6. molten; 7. olivine; 8. shield; 9. sill; 10. volcano

## 5-5 Volcanic Rock Observation

The volcanic rock samples listed are easily attainable in most areas. There are other volcanic rock specimens that may be used: peridotite, diorite, andesite, volcanic breccia, and syenite.

Student observation will vary according to the rocks selected.

## 5-6 Specific Gravity of Volcanic Rocks

Some students will need help dividing decimals. Remind them to take careful readings with the spring balance. If they have trouble holding the balance steady, have them tie the balance to a ring stand.

### *Part 1*

Recording chart measurements will vary.

### *Part 2*

The basaltic rocks, because they have a higher concentration of iron

## 5-7 Build a Model Volcano

### *Part 1*

The flow pattern around the "soda cone" doesn't show anything strange or unique. It simply demonstrates how a liquid substance would flow down the sides of a cone and settle around the base.

### *Part 2*

1. The carbon dioxide gas produced when the vinegar and bicarbonate of soda mixed together
2. A buildup of pressure due to the release of carbon dioxide
3. The buildup of heat and pressure in the Earth's crust

## 5-8 Simulated Volcanic Rock Petroglyphs

The figures and symbols of each student will vary, as will the interpretations of figures and symbols.

You might want to pass along the following information to students: The purpose for studying petroglyphs and their overall importance is to gain knowledge regarding how some early people relied on earth materials for writing messages and record keeping. Petroglyphs are important because information gleaned from them can be compared to historical evidence supplied by artifacts, fossils, ancient art, and primitive writing such as Egyptian hieroglyphics.

# Section 5: Volcanoes
## MINI-ACTIVITIES

Listed below are eight mini-activities, one to five minutes in length, for students to do at the beginning or end of the period.

1. In the squares below, fill in the letters that describe three different lava rocks. A hint is provided for each rock.

| | | | | | | | | | |
|---|---|---|---|---|---|---|---|---|---|
| (basalt) | | | | | | | | | Rhymes with *assault* |
| (pumice) | | | | | | | | | Has a name for a group of animals in the term |
| (obsidian) | | | | | | | | | Has a man's nickname in the term |

2. Locate and darken the squares in the box that contain the letters of a volcanic rock. *Hint:* It is the most common lava; dark, fine-grained, and heavy.

| | | | | | |
|---|---|---|---|---|---|
| c | x | a | j | e | s |
| n | o | e | d | y | p |
| t | v | i | a | m | h |
| w | b | m | c | l | u |
| g | k | r | e | f | c |
| z | c | i | f | n | k |

(**Answer:** basalt)

3. In one or two short sentences, tell how the following statement relates to volcanoes: "Alice, don't vent your anger toward me."

   (**Answer:** A volcano releases its heat and pressure through a vent or opening.)

4. Sketch a volcanic rock using the following description: Dark, coarse-grained texture; various sized crystals scattered throughout the rock. Then answer this question: Did this rock come from fast- or slow-cooling lava? How can you tell?

   (**Answer:** Sketches will vary. The rock probably came from slow-cooling lava because it has a coarse-grained texture and various sized crystals.)

5. What is wrong with this picture?

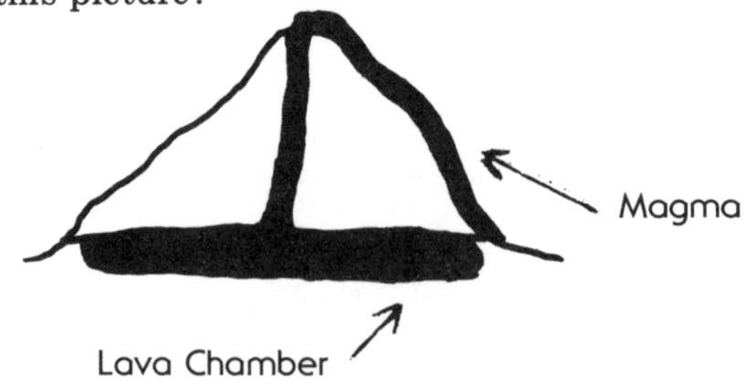


(**Answer:** The labels should read *magma chamber* and *lava.* In other words, *lava* should be the top label and *magma chamber* the bottom label.)

6. Unscramble the names of the following types of volcanoes. They all contain two words. Arrange the answers in alphabetical order.

   iedslh osveolnca; cenos inecrd; oectoimps onsce

   1. ______________________________ (cinder cones)
   2. ______________________________ (composite cones)
   3. ______________________________ (shield volcanoes)

7. See how many people's names, including nicknames, you can make from the following three words by switching the letters around. You can use a letter more than once.
   - Volcano (Cal, Anna, Val, Lana, Al, Lon, etc.)
   - Obsidian (Dan, Anna, Sid, Bob, Ada, Bo, Di, etc.)
   - Composite (Tom, Pete, Moe, Tim, etc.)

8. List five volcanic structures that have the letter *i* in their names.

   (**Answer:** batholith, laccolith, sill, dike, cinder, basaltic magma, felsite, shield cone volcano, etc.)

## Section 6

# EARTHQUAKES

## OUTLINE

- 6-1 Earthquake Terms
- 6-2 Earthquakes Everywhere!
- 6-3 Earthquake Word Search
- 6-4 Think About It
- 6-5 Movement Along a Fault
- 6-6 Moving Blocks
- 6-7 Making Waves
- 6-8 Locating the Epicenter

## MATERIAL

The following laboratory materials are needed for Section 6:

- clay
- colored (red or blue) pencils
- compasses
- earthquake maps 1, 2, and 3 (from Activity 6-8)
- pencils and paper
- plastic, glass, or metal trays
- small pebbles
- straight pins
- styrofoam sheets, 6″ by 2″
- water
- wooden blocks, 2″ by 2″
- world maps or globes

Name ______________________________ Date ____________________

# 6-1 Earthquake Terms

Use the clues to complete each word associated with the study of earthquakes. Then unscramble the circled letters to uncover the answer to the mystery question.

1. The face of a steep slope or cliff — _ s _ a _ p _ _ _ t
2. The point at which rocks break and separate; where an earthquake originates — _ o _ (u) _
3. Vibrations caused by an earthquake — _ r _ _ o _ _
4. The energy patterns recorded on a seismograph — _ a _ e ( )
5. An earthquake of minor intensity compared with a major shock; it comes from the same source as a major shock, but occurs later — _ f ( ) e _ _ h _ c _
6. The thick, dense part of the Earth beneath the Earth's crust — _ _ (n) _ l _
7. A fracture in the crust associated with rock movement — _ a _ l _
8. A point directly above the focus of an earthquake — _ p _ c _ n _ e _
9. An instrument used for recording earth tremors — _ e _ s ( ) _ q _ a _ _
10. The process of rocks breaking and moving apart; a change in position — d _ s _ _ a _ _ m _ _ _
11. Backward and forward movement; a shaking or quivering — _ i _ r ( ) _ i o _
12. The strength or degree of force — i _ t _ _ s ( ) _ y
13. The constraining force produced by an earthquake — p _ _ (s) _ _ r _
14. The outer portion or layer of the Earth — _ r _ s _
15. A shaking of the Earth's crust caused by rock movement along a fault — _ a _ _ h _ _ a _ e

**Mystery Question:** A series of giant waves may be generated by earthquakes on the ocean floor. These are known as "tidal waves." What other name do they go by?

_ _ _ _ _ _ _ _

Name ______________________ Date ______________

# 6-2 Earthquakes Everywhere!

An earthquake may occur anywhere on Earth, but most happen in areas of crustal movement associated with mountain-building processes.

The Pacific Coast region provides a shake or two for its inhabitants, especially in California along the San Andreas Fault—a great rift which extends for hundreds of miles along the western border of California.

Earthquakes occur in patterns known as seismic belts. These are well-defined fault zones. Many earthquakes originate in the circumpacific belt made up of young mountain ranges and chains of volcanic mountains.

## PART 1

Use the copy of the world map provided here and a red or blue colored pencil. Place an X on the world map indicating the approximate location of the following cities or countries:

Algeria (northern tip)
Anchorage, Alaska
Bogotá, Colombia (South America)
Costa Rica
Eureka, California
Guatemala, Central America
Iran
Japan
Kamchatka
Lisbon, Portugal
Los Angeles, California
Madrid, Spain
Mexico City, Mexico
New Guinea
New Zealand
Philippines
Quito, Ecuador (South America)
Santiago, Chile (South America)
Singapore
Tibet



## PART 2

Answer the following questions on the back of this sheet:

1. The X's represent epicenters around the world. Look at the pattern formed by the X's. What does the pattern suggest?
2. Why do you think the pattern from Question 1 occurs along coastlines instead of inland?
3. What force(s) do you think keeps these plates on the move?
4. What do you think is the underlying cause of earthquakes along the coastline?
5. We live in a technological society. Why, then, are there still mysteries about the origin of earthquakes?
6. List three positive and three negative points about an earthquake that registers 4.0 on the Richter scale.

# World Map

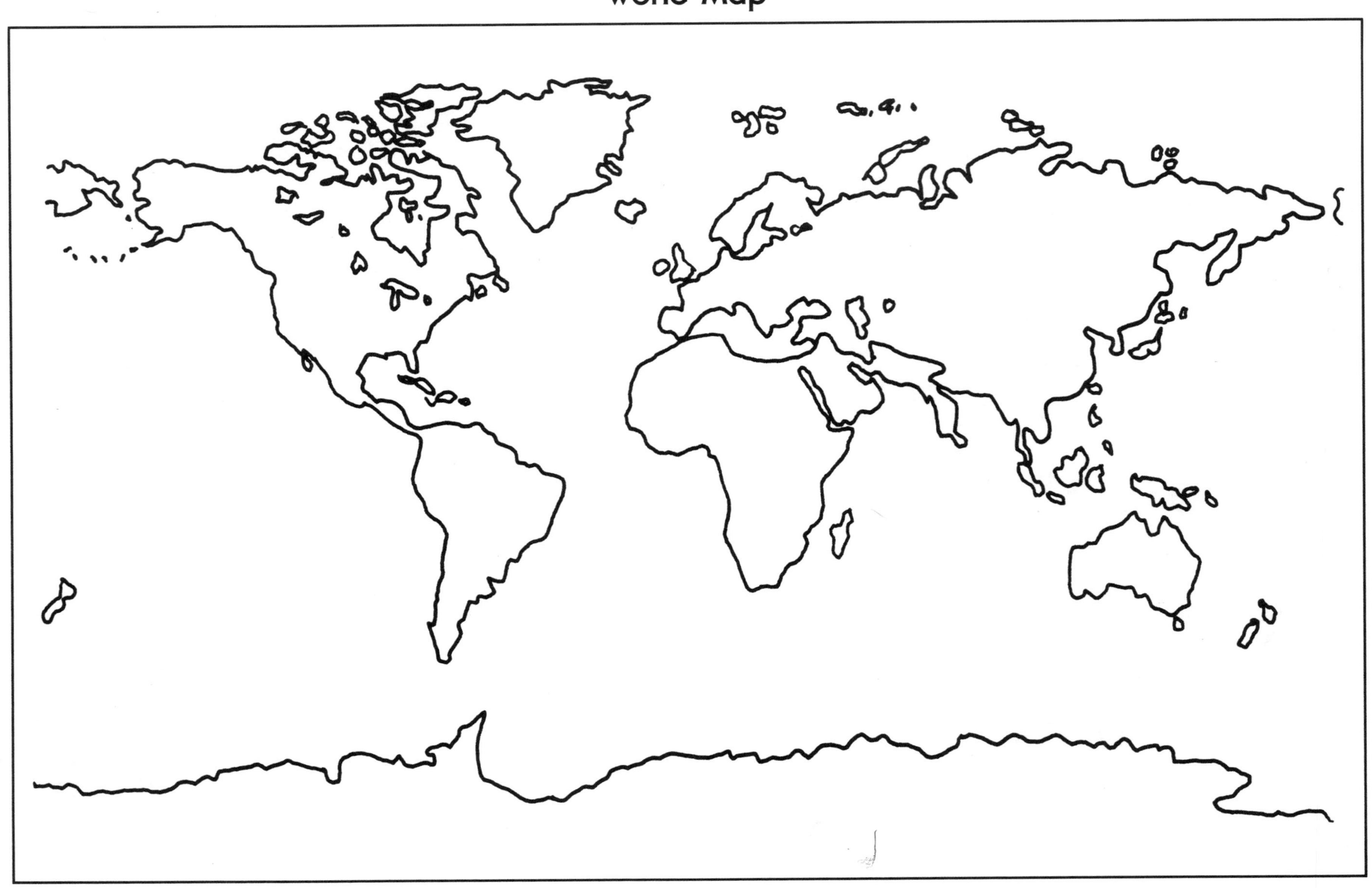

Name ______________________ Date ______________

# 6-3 Earthquake Word Search

Locate and circle the 18 terms associated with earthquakes commonly found in earth science textbooks. The terms are listed below the puzzle. They may be found backward, forward, vertically, horizontally, and diagonally.

```
A E K C O H S R E T F A J B S
S T R A I N Y C N U M G I E E
F H R C I M S I E S Z T C P I
Q L D R S C X B R K A O I I S
K O I P O W V J G F N M B C M
N L S R I M T I Y D A V W E O
S H P G E O E D A N K E X N G
Y O L K Q I J R U I F E W T R
T O A N I L Y S T A M R A E A
I H C P E W T C U F A U B R T
S U E R A B Y L A C I T R E V
N T M V F C T A I E A C S G E
E F E P R I M A R Y W A V E T
T I N A S E I S M O G R A P H
N R T S U R C T E S W F O E Z
I Z M L A T N O Z I R O H A I
```

| | | |
|---|---|---|
| aftershock | fracture | seismograph |
| crust | horizontal | shake |
| displacement | intensity | strain |
| energy | primary wave | tremor |
| epicenter | secondary wave | tsunami |
| fault | seismic | vertical |

**Name** ______________________ **Date** ______________________

# 6-4 Think About It

Whenever an earthquake of sizable proportion occurs, newspapers print the story. A detailed news article usually includes photos of destruction and interviews with eye witnesses and people suffering personal and property losses.

Read the following fictitious newspaper clipping describing an earthquake.

*A monster earthquake ripped through Jamesburg early Monday morning killing 22 people and injuring more than 200. Buildings tumbled to the ground, freeways buckled, and Brighton Dam developed three large cracks at its base. Fires broke out, causing two people to jump out of their high-rise apartments.*

*The quake, registering 6.7 on the Richter scale, sent people scurrying into the streets. The earthquake's focus was centered on the Lester/Richards Fault, six miles east of Jamesburg. Over 20 aftershocks measuring 4 or more on the Richter scale followed the first shock.*

Now answer the following questions regarding the earthquake.

1. Approximately 70,000 people live in Jamesburg. Why do you think only 22 were killed?

2. Many people refuse to leave quake-prone areas for safer ground. Why do you think these people choose to stay?

3. If you lived in Jamesburg, would you consider moving to another area? Why or why not?

4. Suppose the quake struck at noon instead of early morning. Describe the damage you think the earthquake would have done.

5. Why are some people more afraid of aftershocks than the initial earthquake?

6. Do you think insurance companies should sell earthquake insurance to people who live or work near fault lines? Why or why not?

Name ______________________________ Date ______________________

# 6-5 Movement Along a Fault

**Introduction:** A fault is a tear or fracture in the Earth's surface. Movement along a fault may be slightly noticeable or severe enough to cause great damage. The famous San Francisco earthquake in 1906 occurred along a 37-mile stretch of the 700-mile-long San Andreas Fault. Streets cracked open, buildings toppled, and fire destroyed much of the city. Records show that about 700 people died.

Rocks along a fault move sideways (horizontally) or up and down (vertically), sending earthquake waves through the surrounding rock. You can demonstrate this by snapping your thumb and middle finger together. Notice that your thumb moves in an upward direction; your middle finger follows a downward path. The snapping sound indicates a sudden release of energy.

**Objective:** To demonstrate how horizontal or vertical movement along a fault sends earthquake waves through the Earth's crust.

**Materials:** Pencils and paper, styrofoam sheets (6″ by 2″).

## PART 1

**Procedure:** Do the following:

1. Let a styrofoam sheet represent a portion of the Earth's crust. Break the sheet apart by bending the ends downward. What does the cracking sound represent?

   ______________________________________________

2. Fit the broken edges together again. Now separate them slightly. What have you formed from the broken edges?

   ______________________________________________

3. Hold the edges together again. Raise one edge above the other. What type of movement or displacement does this represent?

   ______________________________________________

4. Place both edges together on a desk or tabletop. Slide them back and forth several times. What type of movement or displacement does this represent?

   ______________________________________________

5. Lift the styrofoam pieces from the desk. Notice the small bits of styrofoam left behind. What do these small pieces represent?

   ______________________________________________

## PART 2

Answer these questions:

1. You're in a busy shopping mall during the Christmas rush. An earthquake begins to shake the building. Glass windows break, flower pots overturn, and people begin screaming and running about. What would you do?

2. In your opinion, what would be the worst time for an earthquake to happen in a large, crowded city?

3. Scientists say that San Francisco is overdue for an earthquake of high magnitude (7 or higher on the Richter scale). Why do you think they make this statement?



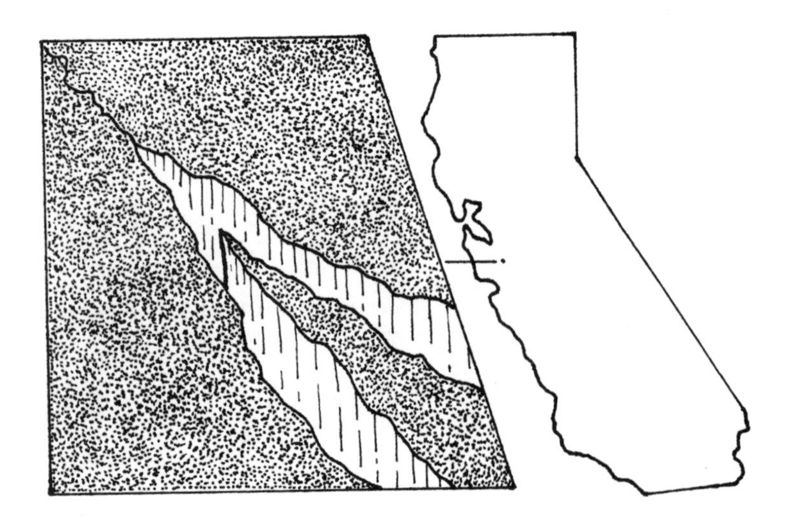

Name ______________________________ Date ____________________

# 6-6 Moving Blocks

**Introduction:** Horizontal or vertical movement along a fault sends earthquake waves through the Earth's crust. These waves spread out in all directions and, depending on their intensity or strength, can be felt many miles away from the earthquake's focus. How much displacement occurs in neighboring rocks is proportional to the location of the epicenter and the makeup of surrounding strata.

**Objective:** To demonstrate the effect of earthquake waves on surrounding rock as they pass through the Earth's crust.

**Materials:** 2″ by 2″ small wooden blocks, straight pins, clay, and pencils and paper.

## PART 1

**Procedure:** Do the following:

1. Set two wooden blocks together on a desk or tabletop. Cover the top of the blocks with a layer of clay. Make a row of straight pins across the length of clay. (Stay to the center of the blocks.)
2. Sketch your model on paper. Pick up the blocks, clay side facing you. Now slowly slide the blocks by moving one block about one-half inch above or below the other block.
3. Make a second sketch.

Answer these questions:

1. How are the sketches similar? ______________________________
2. How are the sketches different? ______________________________

______________________________________________________________

3. What type of motion or displacement does this model show?

______________________________________________________________

## PART 2

Answer these questions on the back of this sheet:

1. How do you think a quake-prone city should prepare people for a possible earthquake?
2. A recent California newspaper article stated, "the state may harbor an unexpected number of invisible, deeply buried, unstable geologic structures also capable of generating an earthquake . . ." What do you think this statement means?
3. What, in your opinion, is the major force that causes crustal blocks to break apart?
4. Earthquakes provide indirect evidence that earth-building processes are going on deep beneath the Earth. What do you think this means?

Name ______________________ Date ______________________

# 6-7 Making Waves

**Introduction:** Rocks under pressure may reach a point where they break apart, snap back in place, and release energy in the form of earthquake waves.

Earthquake or seismic waves spread out in all directions. They vary in size and speed. The following chart lists the three basic types, approximate speeds, and general features.

| **Earthquake Waves** | | |
|---|---|---|
| *Type of Wave* | *Approximate Speed* | *General Features* |
| Primary or P waves | 3.9 miles per second through Earth's crust | Fastest of three waves; detected first on seismograph |
| Secondary or S waves | 2–5 miles per second through Earth's crust | Travel within the Earth, but never along the surface; detected as second set of waves to reach seismograph |
| Long waves or L waves | About 2.2 miles per second; travel near Earth's surface | Last waves to reach seismograph; cause most of the earthquake damage |

**Objective:** To demonstrate how earthquakes release energy in the form of waves. These waves differ in size, speed, and intensity.

**Materials:** Plastic, glass or metal tray, water, small pebbles, and pencil and paper.

**Procedure:** Do the following:

Half fill a tray with water. Set the tray on a table or flat desktop. Strike one end of the tray with your finger.

1. What does your finger striking the tray represent? ______________________

______________________________________________

2. What does your finger motion represent? ______________________

______________________________________________

3. How would you describe primary or P waves in this activity? ______________________

______________________________________________

4. How would you describe secondary or S waves in this activity? ______________________

______________________________________________

5. How would you describe long or L waves in this activity? ______________________

______________________________________________

6. In which direction do the waves move in the tray? ____________________

________________________________________

7. Seismic waves decrease in intensity as they move away from the focus. Why do you think this happens? ______________________________

________________________________________

8. How are these waves different from earthquake waves? ________________

________________________________________

Repeat the procedure, but this time drop different-sized pebbles, one at a time, into the center of the tray.

9. How is dropping pebbles into the tray like tapping the end of the tray with your finger?

________________________________________

________________________________________

10. How is dropping pebbles into the tray different from tapping the end of the tray with your finger? __________________________________

________________________________________

Name ______________________ Date ______________________

# 6-8 Locating the Epicenter

**Introduction:** An earthquake may occur anywhere along a fault line. Rocks may slip within a small area or release energy over several miles. Rock slippage extending over 150 miles occurred along the San Andreas Fault during the San Francisco earthquake.

Seismologists are scientists who study earth movement. They work in seismographic stations throughout the world. When these stations detect earth movement, they compare their records of the time and type of waves and, from that information, determine the epicenter of an earthquake.

Seismograms are the records taken on a seismograph. When an earthquake occurs, seismographic stations at different areas record the disturbance. The seismograms show both the duration and severity of the shock. Seismologists determine the location of the epicenter. If three widely separated seismograph stations report their findings, the location of the epicenter can be detected.

**Objective:** To show how an earthquake's epicenter can be located through seismogram records.

**Materials:** Earthquake maps 1, 2, and 3, compass, pencil and paper.

**Procedure:** Do the following:

1. Examine map 1. There are three seismograph stations—A, B, and C—and three areas marked with an X (labeled A, B, and C).
2. Begin by placing the compass point on Station A. Set the pencil point on the area marked AX. Draw an arc through the AX area. Now move the compass point to Station B, place the pencil point on the area marked BX, and draw an arc through the BX area. Do the same thing for Station C and area CX. Circle the point where all lines meet. This will be the earthquake's epicenter.
3. Repeat the procedure for maps 2 and 3.

Answer these questions:

1. How may seismograph stations come in handy *prior* to a major earthquake?

______________________

______________________

2. How may seismograph stations benefit earthquake victims *after* the damage occurs?

______________________

______________________

3. Do you think seismograph stations should be located near known fault zones only? Why or why not? ______________________

______________________

4. In your opinion, where is the most unlikely place to locate a seismograph station? Why do you think so? ______________________

______________________

# Locating the Epicenter: Map 1

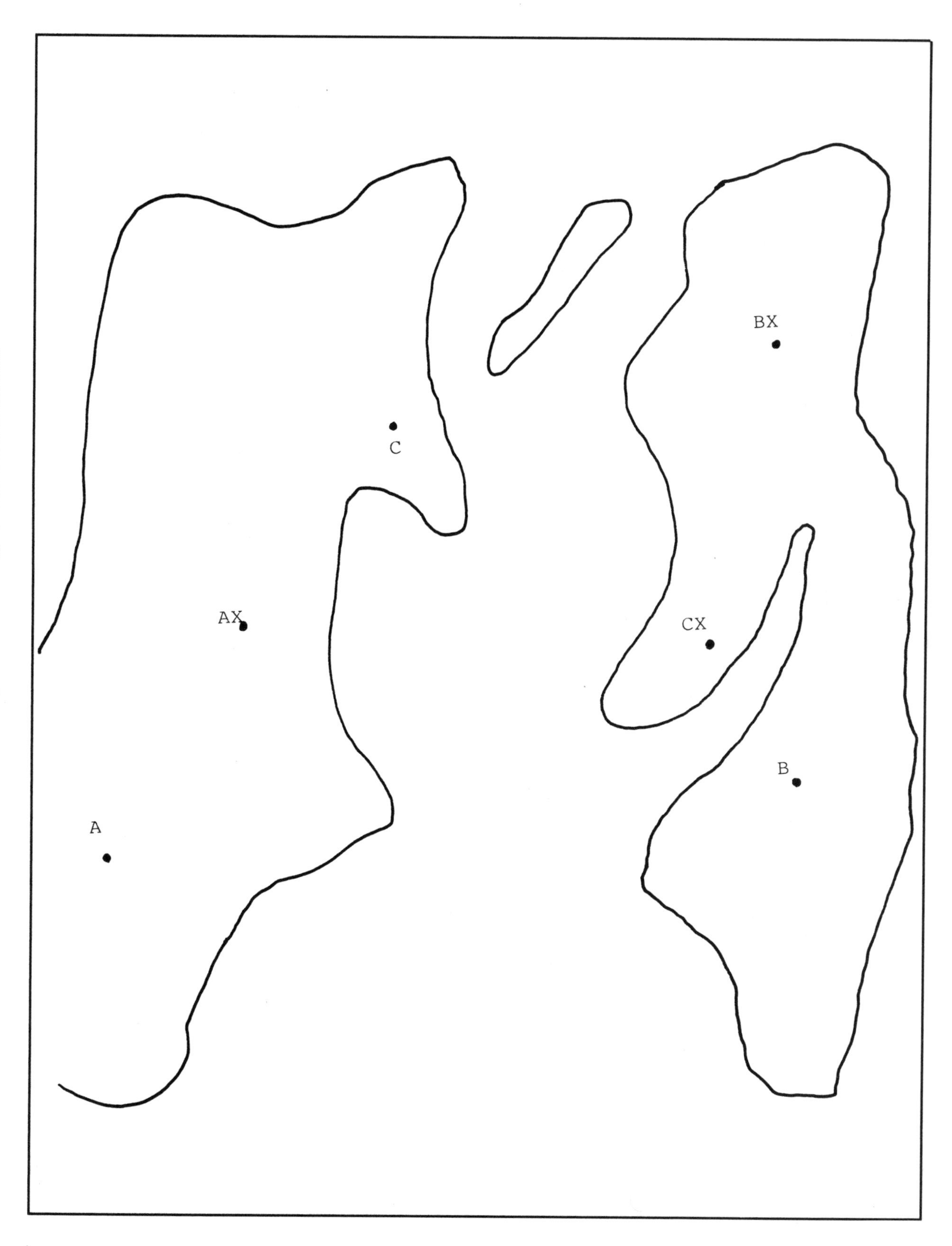

# Locating the Epicenter: Map 2

AX

A

B

C

BX

CX

# Locating the Epicenter: Map 3

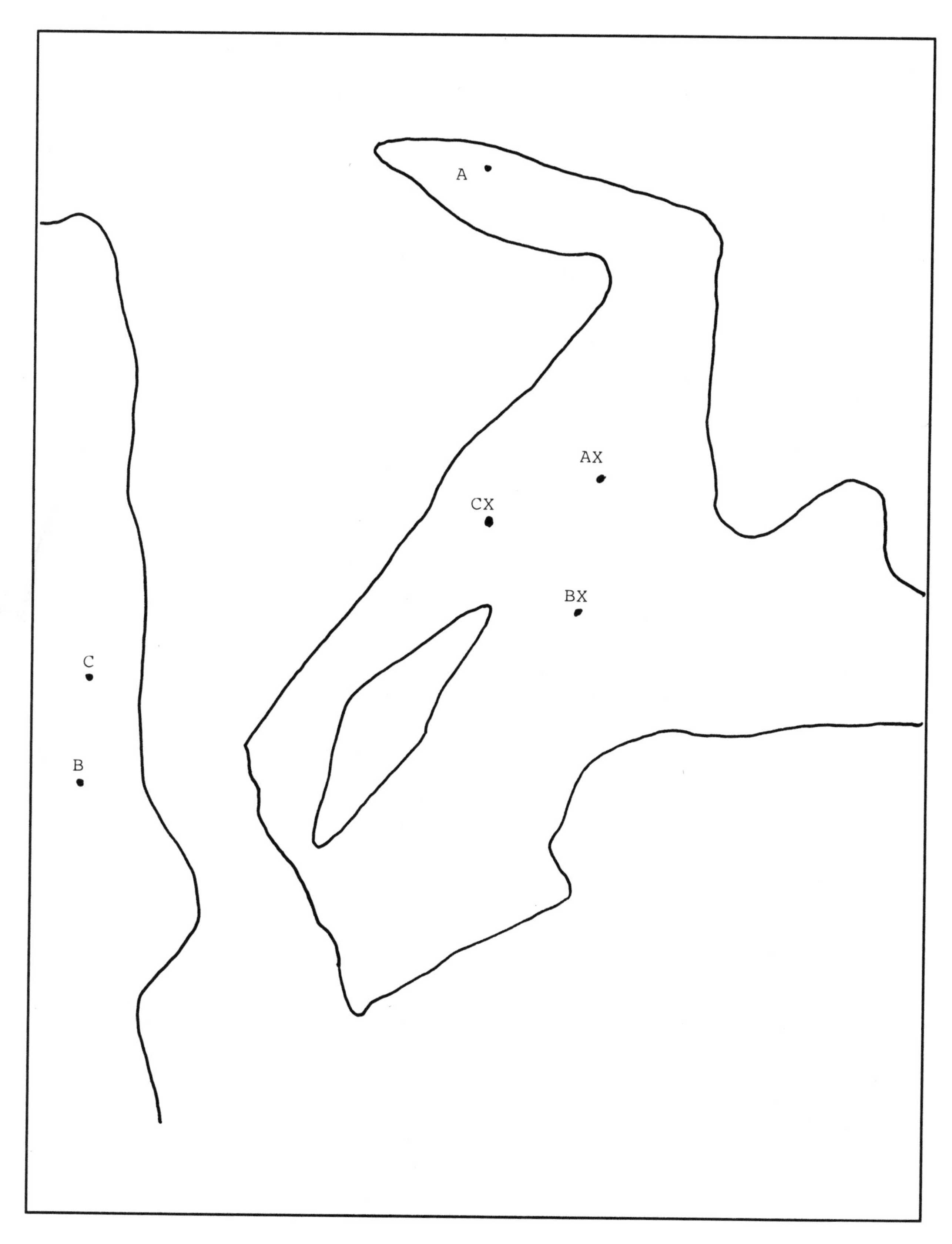

# Section 6: Earthquakes
# TEACHER'S GUIDE AND ANSWER KEY

## 6-1 Earthquake Terms

You might want to mention the term *escarpment* and what it means. Tell students that an escarpment is the face of a slope or cliff. Some cliffs or slopes form from rocks being torn apart by tension.

1. escarpment; 2. focus; 3. tremors; 4. waves; 5. aftershock; 6. mantle; 7. fault; 8. epicenter; 9. seismograph; 10. displacement; 11. vibration; 12. intensity; 13. pressure; 14. crust; 15. earthquake

*Mystery Answer:* tsunamis

## 6-2 Earthquakes Everywhere!

Have enough copies of the world map ready for students. Wall maps of the world, atlases, and world globes will help students locate cities and countries.

***Part 2:***

1. The pattern outlines the borders of major crustal movement.
2. Crustal plates meet near coastline areas; i.e., they mesh together at that point.
3. Heat and pressure originating in the Earth's interior
4. Perhaps two plates meet, crash together, and release energy.
5. No two earthquakes are alike: One may have a focus 40 miles beneath the Earth's surface; another may occur close to the Earth's surface. At best, earthquakes are unpredictable, unique, and enigmatic.
6. *Positive:* a. Only a mild disturbance; b. May give people time to prepare for a bigger shake; c. Allows crustal rocks to relieve strain from the buildup of heat and pressure. *Negative:* a. A frightening event for many people; b. May cause some property damage; c. May create undue panic in crowded areas such as shopping malls.

## 6-3 Earthquake Word Search

| A | E | K | C | O | H | S | R | E | T | F | A | J | B | S |
|---|---|---|---|---|---|---|---|---|---|---|---|---|---|---|
| S | T | R | A | I | N | Y | C | N | U | M | G | I | E | E |
| F | H | R | C | I | M | S | I | E | S | Z | T | C | P | I |
| Q | L | D | R | S | C | X | B | R | K | A | O | I | I | S |
| K | O | I | P | O | W | V | J | G | F | N | M | B | C | M |
| N | L | S | R | I | M | T | I | Y | D | A | V | W | E | O |
| S | H | P | G | E | O | E | D | A | N | K | E | X | N | G |
| Y | O | L | K | Q | I | J | R | U | I | F | E | W | T | R |
| T | O | A | N | I | L | Y | S | T | A | M | R | A | E | A |
| I | H | C | P | E | W | T | C | U | F | A | U | B | R | T |
| S | U | E | R | A | B | Y | L | A | C | I | T | R | E | V |
| N | T | M | V | F | C | T | A | I | E | A | C | S | G | E |
| E | F | E | P | R | I | M | A | R | Y | W | A | V | E | T |
| T | I | N | A | S | E | I | S | M | O | G | R | A | P | H |
| N | R | T | S | U | R | C | T | E | S | W | F | O | E | Z |
| I | Z | M | L | A | T | N | O | Z | I | R | O | H | A | I |

## 6-4 Think About It

Newspaper reporters work hard to generate high interest in catastrophic events. Their stories stimulate the thinking process and open the door for excellent discussion. Use the thought-provoking questions in this activity to encourage students to think and discuss.

You might want to have students collect newspaper stories and write (and answer) their own questions. *Suggestion:* Have students begin their questions in the following manner:

- "What would you do if . . ."
- Start the sentence with *describe, outline,* or *compare.*
- "Do you think . . . (complete the sentence)? Why do you think so?"

These expressions encourage students to think carefully about an issue and require more than a yes or no answer.

1. The quake struck early in the morning. Many people weren't crowding the streets or stores at that time.
2. These people do not want to leave home. They are settled, working in the area, have children going to local schools, etc.
3. Answers will vary.
4. The streets, stores, and other buildings would be crowded with people. There would be far more damage; i.e., death and injuries due to panic-stricken behavior.
5. They fear that a "big one" will strike again soon. Also, the vibrations of shock waves make people feel defenseless, without any protection.
6. Answers will vary.

## 6-5 Movement Along a Fault

You may wish to review the following terms (with diagrams). How many terms you review is up to you. *Fault*—a crack or break in the earth's surface; *fracture*—a break, crack, or split; *energy*—capacity for doing work or the cause of motion; *horizontal*—along a line parallel to the horizon; *vertical*—at a right angle to the horizon; *displacement*—movement or shifting of position, as in rocks sliding past one another

***Part 1:***

1. A sudden release of energy
2. A fault line
3. Vertical displacement
4. Horizontal displacement
5. Pieces of broken rock

***Part 2:***

1. Answers will vary.
2. Rush hour, at lunch time, when schools are in session, or during a crowded event
3. San Francisco lies near several fault zones; it's likely that one of these faults may become active.

## 6-6 Moving Blocks

***Part 1:***

1. Both represent crustal material; i.e., land structures, rock formations, etc.
2. Sketch 1 shows the Earth's crust before an earthquake; sketch 2 shows displacement after an earthquake.
3. Vertical movement or displacement

***Part 2:***

1. By distributing literature on what to do before, during, and after an earthquake; educate school-age children; educate people through the media; etc.
2. Two things: Some earthquakes occur in places other than well-marked, well-mapped faults; and earthquake predictions are very difficult to make.
3. Heat and pressure within the earth.
4. We may not be able to see heat and pressure working on crustal rocks, but the sudden release of energy (earthquakes) indicates that something is going on.

## 6-7 Making Waves

Show students, using diagrams, how the three different waves (P, S, and L) might look on a seismograph.

Primary (P) Secondary (S) Surface (L)

Remind students that most of the damage comes from the slow surface, or L, waves.

1. The point or focus where the earthquake originated.
2. Rocks breaking and snapping back in place.
3. The primary, or P, waves would be the first waves to move across the water in the tray.
4. The secondary, or S, waves would be the second series of waves to move across the water in the tray.
5. The long, or L, waves would be the last waves to move across the tray.
6. Away from the point of finger/tray contact
7. The waves lose energy and strength as they travel through the Earth's crust.
8. They are man-made; earthquake waves originate from natural events and may cause heavy damage.
9. The tapping of the tray and pebbles striking the water release energy in the form of waves; both points of contact simulate the focus of an earthquake.
10. The force and location of contact is different in both situations.

## 6-8 Locating the Epicenter

1. By picking up minor crustal disturbances, they can warn people of a possible big earthquake in a certain area.
2. Victims learn the location of the epicenter. They can prepare themselves accordingly for possible aftershocks.
3. Answers will vary. Students should support the idea of having these stations located near known fault zones *and* suspected fault zone areas.
4. Probably in an area where earthquakes have never been reported; there may not be a fault zone for miles around.

# Section 6: Earthquakes
## MINI-ACTIVITIES

Listed below are seven mini-activities, one to five minutes in length, for students to do at the beginning or end of the period.

1. Complete the puzzle.

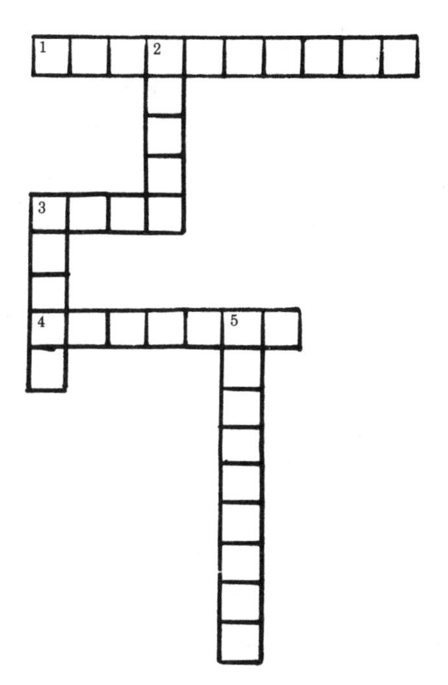

**Across**

1. The study of earthquakes
3. Center of the Earth
4. Characteristics of earthquakes

**Down**

2. Richter developed this
3. Outermost layer of the Earth
5. Strength of an earthquake

(**Answers:** *Across*—1. seismology; 3. core; 4. seismic; *Down*—2. scale; 3. crust; 5. intensity)

2. You have two minutes to write down as many words as you can to describe an earthquake. (Answers will vary.)

3. Unscramble the words that complete the following sentences:

   Most earth (mtsorer) ________________ are so slight that they may not be (tecddeet) ________________. However, they can be recorded on a (gimshpareos) ________________.
   (**Answers:** tremors, detected, seismograph)

4. These three statements are false. Change them into true statements:
   - Albert Einstein developed the Richter scale for measuring earthquake strength.
     (**Answer:** Dr. C. F. Richter developed . . . )
   - The focus of an earthquake is the point on the surface of the Earth above the location where the disturbance has taken place.
     (**Answer:** The epicenter of an earthquake . . . )
   - A seismogram is the instrumental device for recording earth tremors 4.5 or above on the Richter scale.
     (**Answer:** . . . 1.0 or above on the Richter scale)

5. You have five minutes to make a list of eight terms associated with earthquakes from the letters listed in the box. You may use some letters more than once.

| w | s | g | m | y | r | v | l | a | n | e | f | h | t | i | k | u | c |
|---|---|---|---|---|---|---|---|---|---|---|---|---|---|---|---|---|---|

(**Possible answers:** fracture, fault, seismic, crack, shake, wave, Richter scale, energy, etc.)

6. Make-believe book titles and authors are fun to create. In fact, the sillier the better. Try one or two regarding earthquakes. Here are two examples to get you started:
   - *The Restless Earth* by Rock Shakin
   - *Moving Mountain* by Des Plazemint

   (Answers will vary.)

7. Some scientists believe earthquakes are caused by heat and pressure buildup within the Earth's crust. This is associated with massive plate movement, they say. List another reason you think earthquakes occur.

   (Answers will vary.)

## Section 7

# FOSSILS

### OUTLINE

7-1 What Is a Fossil?
7-2 Fossil Term Roundup
7-3 Who Was That Dinosaur?
7-4 Fossil Clues
7-5 A Survey of Bones
7-6 Plaster Fossil Model
7-7 Chalk Stick Fossil Model
7-8 Wax Imprints
7-9 Molds and Casts
7-10 Snail Cast
7-11 Pencil Print
7-12 Forever Amber, Part 1
7-13 Forever Amber, Part 2
7-14 "Reconstruct," a Fossil Game

### MATERIAL

The following laboratory materials are needed for Section 7:

- aluminum foil
- beakers
- burners
- candle wax
- clay
- colored chalk sticks
- colored crayons
- dead insects
- dissecting needles
- dried garden snail shells
- food coloring
- fossil diagrams or fossil models
- goggles and aprons
- index cards (5″ by 7″)
- jar lids
- Karo syrup
- leaves
- metal pans
- molasses
- mucilage (amber colored) for paperwork
- nails
- newspaper
- paperclips
- paper towels
- pencils
- petroleum jelly
- plaster of Paris
- rulers
- scissors
- seashells
- shallow trays
- stirrers
- styrofoam cups
- sugar
- tongs
- tweezers
- twigs
- watches
- water
- wire

Name ______________________ Date ______________________

# 7-1 What Is a Fossil?

## PART 1

Below are 10 statements regarding fossils. Place a check mark (✓) in the space to the left of each true statement. (*Hint:* There are six true statements.)

___ 1. All organisms die and become fossils.

___ 2. A fossil may be an impression of the original organism.

___ 3. Many fossil records are molds and casts.

___ 4. Organisms with soft parts have the best chance of becoming fossils.

___ 5. Some organisms leave behind traces of carbon as fossil evidence.

___ 6. Permineralization occurs when minerals replace original hard parts of an organism.

___ 7. Tar pools trapped and preserved animals that lived during the Precambrian era.

___ 8. Ancient insects, trapped in tree resin, left behind well-preserved body structures.

___ 9. An organism may die, dry out, and leave behind mummified remains.

___ 10. The process of metamorphism seldom distorts or destroys fossil evidence.

Now write a definition for *fossil* using the following key terms: *life, evidence, past,* and *preserved.*

A fossil is ______________________________________________

______________________________________________

______________________________________________

______________________________________________

______________________________________________

______________________________________________

## PART 2

Check over the false statements. Change them into true statements by adding or deleting words or by rearranging the sentences. Write your rewritten statements on the back of this sheet.

## PART 3

Some fossils are more common than others. The left-hand column lists four common fossils. In the space at the right, make a sketch to show what each fossil looks like.

| Fossil | Sketch |
| --- | --- |
| Crinoid | |
| Trilobite | |
| Plant leaf | |
| Ammonite | |

Name ______________________ Date ______________________

# 7-2 Fossil Term Roundup

## PART 1

A term is missing in each sentence below. Fill in each missing term. Then solve the puzzle by answering Question 10.

1. An _ _✓ _ _ _✓ _ _ is a mark or pattern made by pressure.

2. A _✓ _ _ _ _ _ _ _ _ is an animal that feeds on dead organisms.

3. _ _ _ _ _✓ _ _ _✓ _ _ _ _ _ is a process in which an organism becomes partly decomposed and leaves a residue of carbon.

4. A fossil gum from the sap of ancient plants is known as _ _ _ _✓ _.

5. _ _ _ _ _ _ _ _ is the material deposited by water or air.

6. Shale is an example of sedimentary _ _ _✓ _.

7. Many _ _ _ _ _ remains from the Pennsylvanian period provide the coal source we use today.

8. A _ _✓ _ _ is a cavity or depression left by an organism.

9. The filling material of a mold—dirt, sand, or mud—is called the _ _ _ _.

10. In what geologic era did the dinosaurs dominate the Earth? You must unscramble the checked letters to discover the answer. ______________

## PART 2

There are 12 organisms listed in the puzzle that have been found as fossils. Find the organisms and circle them. Then write each name (in any order) on the numbered spaces below the puzzle. Answers may be found forward, backward, horizontally, vertically, and diagonally.

| | | | | | | | | | |
|---|---|---|---|---|---|---|---|---|---|
| i | a | e | i | o | s | a | l | i | a |
| b | l | u | a | p | t | v | n | b | u |
| c | f | l | o | w | e | r | e | k | e |
| n | j | n | y | c | b | i | h | w | i |
| r | g | e | o | f | l | s | f | c | o |
| e | m | r | o | w | i | a | p | i | a |
| f | a | l | z | f | h | s | m | w | s |
| l | k | i | r | i | q | b | h | o | l |
| a | s | a | t | s | c | y | c | a | d |
| e | t | n | a | h | l | p | i | k | n |
| s | h | s | i | f | y | l | l | e | j |

1. ______________________

2. ______________________

3. ______________________

4. ______________________

5. ______________________

6. ______________________

7. ______________________

8. ______________________

9. ______________________

10. ______________________

11. ______________________

12. ______________________

Name ______________________________ Date ______________________

# 7-3 Who Was That Dinosaur?

Dinosaurs ruled the Earth during the Mesozoic era. Some dinosaurs, like the brontosaur and tyrannosaurus, have been glorified in the movies as murderous reptiles able to crush anything in their paths.

## PART 1

How well do you know dinosaurs? Match the partial sketches of five different dinosaurs in the right-hand column with the dinosaur name in the left-hand column. Write the letter of the sketch in the space to the left of the number. Put an X in the two spaces that do not apply.

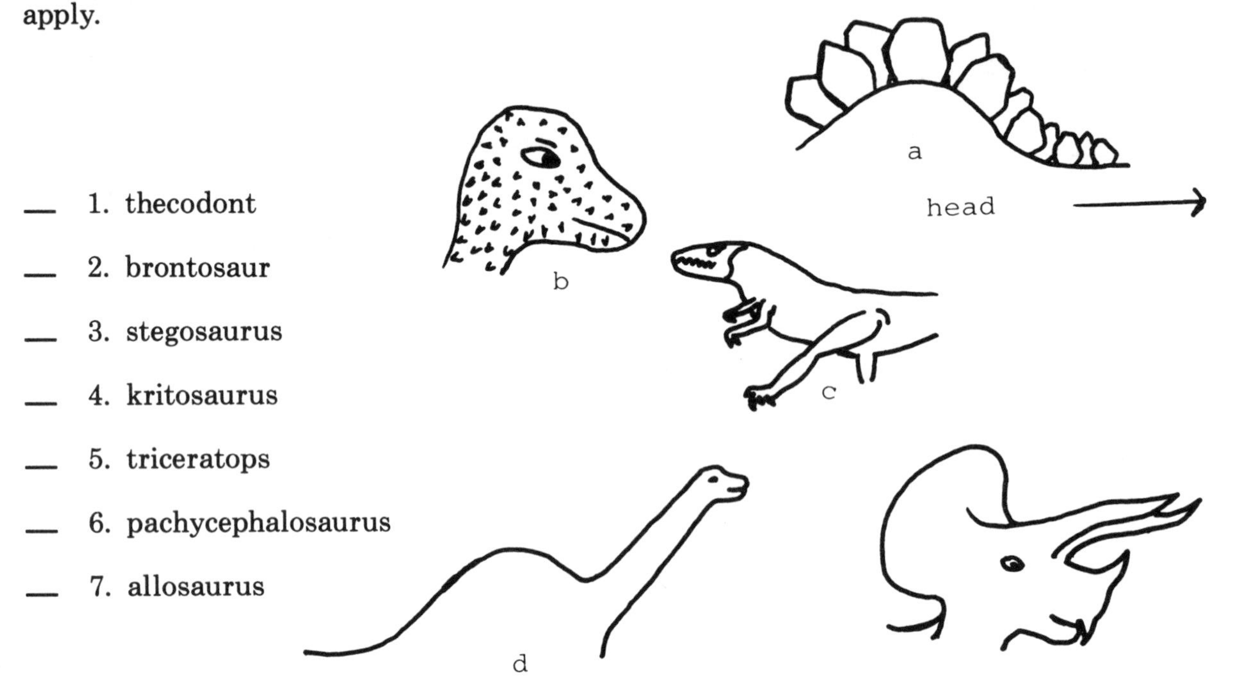


__ 1. thecodont

__ 2. brontosaur

__ 3. stegosaurus

__ 4. kritosaurus

__ 5. triceratops

__ 6. pachycephalosaurus

__ 7. allosaurus



## PART 2

Match the dinosaur name listed in the left-hand column with its description in the right-hand column. Draw a line to connect the name with the description.

| | |
|---|---|
| 1. stegosaurus | a. known as "thick-headed" reptile |
| 2. brontosaur | b. famous for "second brain" |
| 3. pachycephalosaurus | c. largest of "lizard-hipped" dinosaurs |
| 4. allosaurus | d. "terrible teeth" |
| 5. thecodont | e. "socket teeth" |
| 6. triceratops | f. "horned dinosaur" |

Name ______________________________ Date ______________________

# 7-4 Fossil Clues

Write the answers to the numbered clues in the space provided. These answers will give you the words to complete the crossword puzzle. The crossword contains letter clues to help you place the words.

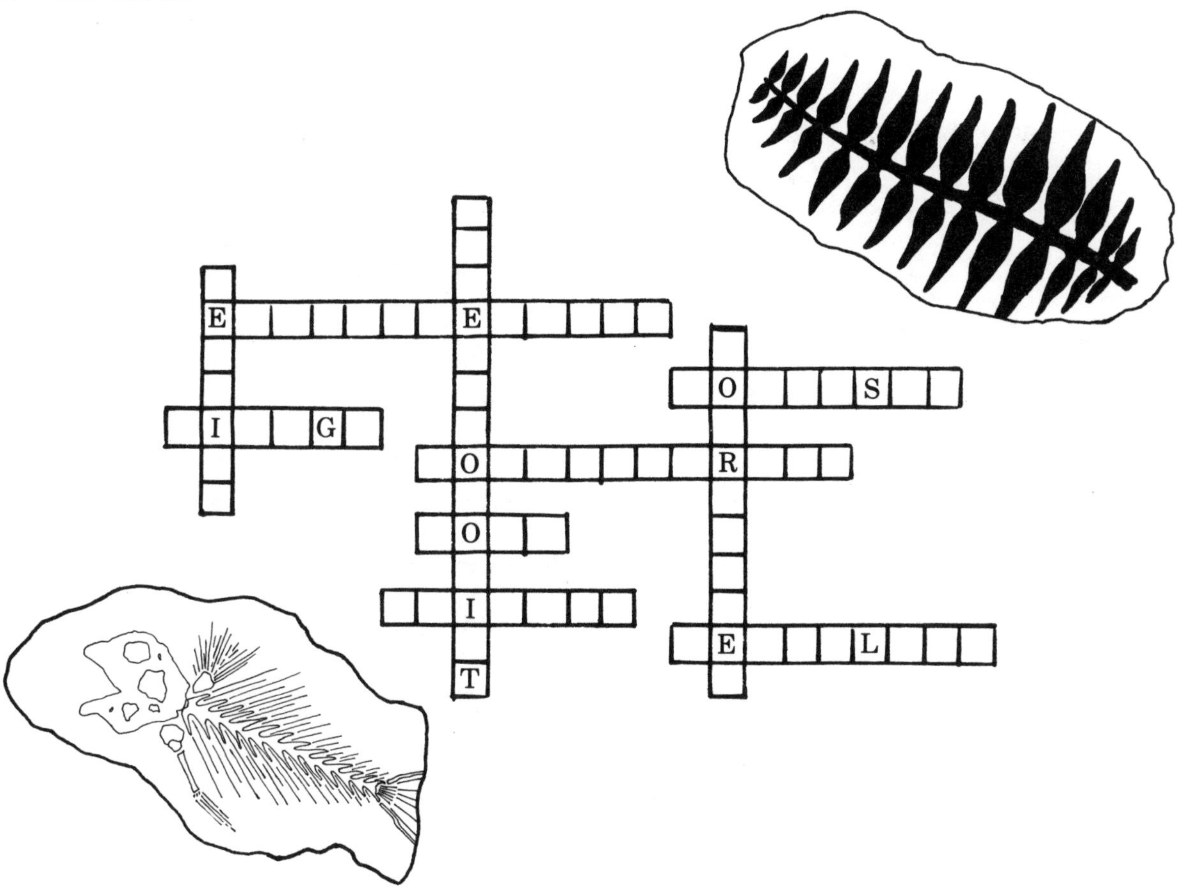




1. A scientist who studies fossils: ______________________________
2. Term meaning to "turn to stone": ______________________________
3. A ________________ print is indirect evidence of past life.
4. Fossil waste material or excrement: ______________________________
5. Fossil corals belong to the phylum: ______________________________
6. Fossil crinoids or sea lilies belong to the phylum: ______________________
7. Brachiopods are a phylum of marine shelled: ______________________
8. The word ________________ means "soft bodied."
9. Archaeopteryx is a fossil animal that shared both ________________ and bird traits.
10. An example of a "living plant fossil" is a ______________________________.

Name ______________________ Date ______________________

# 7-5 A Survey of Bones

In general, the hard parts of an organism stand the best chance of becoming fossilized, especially if they undergo rapid burial. Bones, teeth, shells, and plant stems provide excellent examples of hard parts turning into fossils.

## PART 1

Use a picture of the human skeleton or a model skeleton to do the following:

1. List four structures or parts of the human skeleton that you feel are the hardest.

   a. ______________________ b. ______________________

   c. ______________________ d. ______________________

2. List four structures or parts of the human skeleton that you feel lie somewhere between hard and soft.

   a. ______________________ b. ______________________

   c. ______________________ d. ______________________

3. List three structures or parts of the human skeleton that you feel are the softest.

   a. ______________________ b. ______________________

   c. ______________________



## PART 2

Answer these questions on the back of this sheet:

1. What parts of a fossilized human skeleton are usually preserved?
2. Why do the parts listed in Question 1 become fossilized?
3. Why do mummified human remains usually have sunken chest cavities?
4. Why do you think cotton pads are stuffed between vertebrae on a human skeleton?
5. List other places on the human skeleton where synthetic material has replaced lost parts.
6. What conditions might cause a hard-bodied organism to become fossilized?
7. What conditions could cause a hard-bodied organism not to be fossilized?
8. Would you like your remains to become fossilized? Why or why not?
9. What could future generations learn from studying your fossilized remains?
10. How could the study of fossilized human remains improve living conditions on Earth?
11. One-celled organisms, worms, sponges, and jellyfish all have soft body parts. Why do you think some of them become fossilized?

Name ______________________________ Date ____________________

# 7-6 Plaster Fossil Model

**Introduction:** Fossils are the remains or traces of organisms that lived during ancient geological times. Simply, a fossil is any preserved evidence of past life. Ancient geological times refer to any organism, plant or animal, that is 10,000 years old or older.

Some organisms become fossils through the sedimentation process. A plant or animal dies and becomes buried in sediment like mud or sand. In time the sediment hardens, encases the organism, and preserves it. If the organism doesn't decay or become mangled or eaten by scavengers, it may turn into a fossil.

**Objective:** To make a model impression of an organism in sedimentary-like material.

**Materials:** Fossil diagrams or fossil models, 5″ by 7″ index card, plaster of Paris, pencil, newspaper, water, stirrer, styrofoam cup, food coloring, jar lid, paperclip, nail or dissecting needle.

**Procedure:** Do the following:

1. Select an organism from a fossil diagram or model to represent the plaster model you will make. Stay away from large animals—dinosaurs, whales, elephants, and so on. Select one that is proportionate to the size of the jar lid; for example, insect, leaf, clam, snail, trilobite, and so on.

2. Spread a newspaper over your work area.

3. Mix plaster with water and add two or three drops of food coloring (if desired) in a styrofoam cup. Stir the mixture until a thick paste forms. *Note:* Slowly pour the water into the plaster, stir, and, when the mixture reaches a milkshake consistency, stop stirring. Tap the bottom of the cup against a desk or table to help smooth out the mixture.

4. Pour the mixture into a jar lid. Again, tap the jar lid to help the mixture settle.

5. Set the mixture aside to dry.

6. When the mixture is dry, begin making the model by sketching the outline of the selected organism on the plaster.

7. Use a nail, dissecting needle, or straightened-out paperclip to carve the model fossil.

8. After finishing the model, take an index card and fill out your name, period, and organism information. (See example.)

On the back of the card, briefly describe the organism—what it looked like, what it ate, how it lived, and so on.

Vasquez, Sylvia
Period 3

Fossil model: trilobite
Type of fossilization:
sedimentation
Organism lived during the
Paleozoic era
Habitat: marine environment;
shallow seas

Name ______________________________ Date ____________________

# 7-7 Chalk Stick Fossil Model

**Introduction:** An organism may die, become buried, and leave an impression or indentation in rock. While the organism itself has long vanished, it has left behind evidence of ancient life. Ferns, insects, and worms provide excellent imprints.

**Objective:** To make a model impression of the original organism.

**Materials:** Colored chalk sticks, paperclips, scissors, newspaper, dissecting needle, and fossil diagrams or fossil models.

**Procedure:** Do the following:

1. Place newspaper over your work area.
2. Select an organism from fossil diagrams or models. Ferns, insects, and worms make suitable models.
3. Make a smooth, flat surface on the chalk by scraping it with a scissor blade. *Note:* School chalk is made from the mineral gypsum and contains no fossils.
4. Straighten a paperclip and use it to scratch a fossil figure on the chalk.
5. Use a dissecting needle to bring out detail in your model.

Answer the following questions:

1. What does the chalk represent? ______________________________

______________________________

2. Why do you think a large number of organisms become preserved in natural chalk?

______________________________

______________________________

______________________________

______________________________

3. There are no fossils in the chalk used by your teacher, yet fossils may be found in natural chalk. Explain the meaning of this statement.

______________________________

______________________________

______________________________

______________________________

______________________________

Name ________________________________ Date ________________________

# 7-8 Wax Imprints

**Introduction:** Many fossils are preserved in sedimentary rock. Sedimentary rock is material composed of fragments of other rocks deposited after transportation from their sources. Sandstone, limestone, and shale are examples.

**Objective:** To make a model impression of the original organism.

**Materials:** Candle wax, food coloring, colored crayons, beaker, burner, newspaper, tongs, twigs, leaves, shells, clay, safety goggles, and apron.

**Procedure:** Do the following:

1. Spread newspaper over your work area.
2. Flatten out a half-inch-thick layer of clay.
3. Press a twig, leaf, or shell into the clay pad.
4. Melt wax slowly in a beaker. Add a colored crayon (wrapper removed) or food coloring to the wax. Dark-colored food coloring works best and brings out finer detail.
5. Remove the beaker from the burner with tongs. Pour molten wax over the specimen in the clay. Be sure to use enough wax to cover the entire specimen.
6. When the wax cools, carefully remove the clay and specimen.

Answer these questions:

1. Why do you think few fossil imprints are found in coarse-grained, sedimentary materials?

   ______________________________________________

   ______________________________________________

   ______________________________________________

2. Imprints are not found in every sedimentary layer of rock. Why do you suppose this is true?

   ______________________________________________

   ______________________________________________

   ______________________________________________

3. Which one of the following organisms has the best chance of becoming fossilized: worm, bacteria, fig, jellyfish, or slug? Why?

   ______________________________________________

   ______________________________________________

   ______________________________________________

Name ______________________ Date ______________

# 7-9 Molds and Casts

**Introduction:** Shells leave excellent fossil molds. A mold is a cavity or depression left by an organism. When material such as dirt, sand, or mud fills the mold, the filling material is called a cast.

**Objective:** To make model fossil molds and casts.

**Material:** Modeling clay, seashells, wax, crayons, petroleum jelly, beaker, burner, stirrer, safety goggles, and apron.

**Procedure:** The procedure is broken down into two parts—Part 1, Molds; and Part 2, Casts.

## PART 1: MOLDS

Do the following:

1. Roll out a half-inch-thick layer of clay.
2. Press two or three seashells into the clay.
3. Remove shells and set clay mold aside for Part 2.

Answer these questions:

1. Other than scavengers and destructive bacteria, what might cause an organism to disappear after burial? ______________________

______________________

2. The material that fills a mold comes from an organism's environment. Why, then, is a cast considered a fossil? ______________________

______________________

3. What does the removal of seashells represent? ______________________

______________________

4. List three events that simulate how a mold forms in nature.

______________________

______________________

5. How do fossil molds help us understand the Earth's geological history?

______________________

______________________

## PART 2: CASTS

Do the following:

1. Place wax in a beaker. Melt the wax over a burner. Do not boil the wax. Add a small piece of crayon (color optional).
2. Stir until the crayon and wax mix together. Let the wax cool for approximately five minutes.
3. Apply a thin film of petroleum jelly to each clay imprint. This keeps the wax from sticking to the clay.
4. Pour the wax into the clay imprints. Allow the wax to harden before removing the fossils.

Answer these questions:

1. What earth conditions would allow cast material to duplicate a seashell's external features?

______________________________________________________________

______________________________________________________________

2. What might happen to a mold if the cast material dissolves or wears away?

______________________________________________________________

______________________________________________________________

3. What major features might casts reveal? ______________________________

______________________________________________________________

4. What materials beside sand, dirt, and mud might form casts?

______________________________________________________________

______________________________________________________________

**Review**

A mold is a cavity or impression left by an organism that has decayed or dissolved. A cast is the filling of a mold with a substance that duplicates the shape and texture of the object.

Name ______________________________ Date ____________________

# 7-10 Snail Cast

**Introduction:** Refer to Activity 7-9: Molds and Casts.

**Objective:** To make a model of a fossil cast.

**Materials:** Dried garden snail shells, wire, paper towels, clay, tweezers, colored crayons, burner, beaker tongs, and beaker.

**Procedure:** Do the following:

1. Clean out a dry snail shell with wire.
2. Remove the paper crayon wrapper. Put the crayon in a beaker and place the beaker over a burner.
3. Slowly melt the crayon over a low flame.
4. Use tongs to remove the beaker from the burner.
5. Make a clay pad to hold the snail steady. Gently push the snail shell into the clay pad.
6. Pour the melted crayon into the empty shell. Set the shell aside and let the wax harden.
7. After the wax hardens, remove the snail shell by peeling it away with your fingernail or tweezers.
8. If you have time, make several casts. Try for a "perfect" cast—one that is free of air holes, wrinkles, or rough spots.

Answer the following questions:

1. What does the snail shell represent? ______________________________
2. What does the crayon represent? ______________________________
3. List two materials found in the Earth's crust that could easily fill a snail shell.

   ______________________________

   ______________________________

4. What could you learn about a modern-day snail by studying its cast? ______________

   ______________________________

5. What might you learn from studying a fossil snail shell cast? ______________

   ______________________________

   ______________________________

Name ______________________________ Date ______________________

# 7-11 Pencil Print

**Introduction:** Some leaves, fish, worms, and shrimp leave a carbonized film on rocks as fossil evidence. As the organism decays, enough volatile or organic materials are left behind to form a thin film of carbon. Simply, protoplasm (all the living soft material that makes up an organism) consists of four simple elements—carbon, hydrogen, oxygen, and nitrogen. Carbon makes up about 10.5% of protoplasm. Therefore, once an organism loses nitrogen, oxygen, and hydrogen, the remaining carbon leaves only a trace of the original specimen.

**Objective:** To make a model fossil of a carbonized organism.

**Materials:** Index cards, leaves, and a pencil.

**Procedure:** Do the following:

1. Place a leaf under the top half of an index card.
2. Move a pencil back and forth several times across the index card. Apply firm and steady pressure. Center the movement directly above the leaf.
3. Keep the pencil moving until the leaf pattern appears on the index card. *Note:* The pencil lead simulates the carbon residue left behind by the organism.
4. Clean the print by erasing pencil markings that extend past the leaf imprint.
5. Write *plant leaf* under the print. Describe in two or three sentences how an organism can become a carbonized fossil.

Answer these questions:

1. What does the index card represent? ______________________________
2. What does the leaf represent? ______________________________
3. What does the moving pencil represent? ______________________________

______________________________________________________________

4. What does the pencil lead represent? ______________________________

______________________________________________________________

Name ______________________________ Date ____________________

# 7-12 Forever Amber, Part 1

**Introduction:** Amber, a fossil gum from the sap of ancient plants, trapped and preserved insects and spiders. After the gum-like resin hardened, the insect or spider dried to almost nothing. Certain structures such as heads, antennae, wings, bodies, and legs were well preserved. Several complete organisms have been found preserved in amber.

**Objective:** To make a model fossil of an insect trapped in amber.

**Materials:** Dead insects, sugar, Karo syrup, metal pan, stirrer, water, molasses, shallow tray, burner, safety goggles, and apron.

**Procedure:** Do the following:

1. Mix two parts sugar, one part Karo syrup, and one part water in a beaker.
2. Boil the solution in a pan for about 10 minutes. *Note:* Continue to test the solution until it reaches the right "cracking point"—that is, it becomes brittle when dipped in cool water. Add molasses, one drop at a time, until the desired yellowish-amber color is reached. CAUTION: Be aware of possible beaker breakage.
3. Remove the ingredients from the burner. Pour into a shallow tray.
4. Drop several dead insects into the substance. Arrange insects in different positions throughout the solution. When the solution cools, it hardens and encases the insects. Trapped air bubbles which surround the insects make the amber setting appear real.



Answer these questions:

1. Why do you think insects were attracted to the tree resin?

______________________________________________________________

______________________________________________________________

2. How was the resin able to keep the insects from decomposing?

______________________________________________________________

______________________________________________________________

3. How are insects sealed in amber similar to the animals trapped in tar? How are they different?

______________________________________________________________

______________________________________________________________

4. Why might scientists learn more from studying insects preserved in amber than from studying imprints?

______________________________________________________________

______________________________________________________________

Name ______________________________ Date ______________________

# 7-13 Forever Amber, Part 2

**Introduction:** Refer to Activity 7-12: Forever Amber, Part 1.

**Objective:** To make a model fossil of an insect trapped in amber.

**Materials:** Dead insects, aluminum foil, and mucilage (amber-colored) for paperwork.

**Procedure:** Do the following:

1. Make a shallow tray out of aluminum foil (2″ by 2″ by 1″).
2. Slowly pour a quarter-inch-high layer of mucilage into the tray. Add two or three dead insects.
3. Set the tray aside to dry. (Allow two or three days for drying.)
4. Pour another quarter-inch-high layer of mucilage into the tray.
5. Set aside to dry for two or three days.
6. Pour another half-inch-high layer of mucilage into the tray.
7. Set aside to dry for four, five, or more days.
8. Inspect the tray. You should have a clear view of each insect.

Answer these questions:

1. How is the fossilization by amber process different from the simulation or model version?

______________________________

______________________________

______________________________

______________________________

2. How are the two processes alike? ______________________________

______________________________

______________________________

______________________________

3. Do you think the mucilage can preserve insects for long periods of time? Why or why not?

______________________________

______________________________

______________________________

______________________________

Name ______________________________ Date ______________________

# 7-14 "Reconstruct," a Fossil Game

**Introduction:** Fossils buried in the Earth's crust do not always rest in peace. A fossil encased in sedimentary rock may undergo a significant change. Sedimentary rock, for example, may change into metamorphic rock.

Metamorphic rocks are rocks that have changed by heat, or pressure, or both. For instance, heat or pressure exerted on the sedimentary rock limestone can change it into the metamorphic rock marble. The heat and pressure would distort or completely destroy any fossil evidence. Therefore, few recognizable fossils are found in metamorphic rock due to the tremendous forces which flatten, squeeze, or stretch fossils beyond recognition.

**Objective:** To experience the difficulty of assembling a distorted "organism."

**Materials:** Paper, pencil, ruler, watch, and scissors.

**Procedure:** Do the following:

1. Sketch an animal or plant organism on paper. The sketch should be a simple outline with few details and cover at least a half page. Examples: clam, snail, fish, leaf, fern, and so forth.
2. Cut out the figure. Now cut the figure into 8 or 10 jigsaw puzzle-like pieces.
3. Find a partner who also drew an organism and cut out the pieces. Exchange puzzle pieces with your partner.
4. Take turns putting puzzles together. For example, Partner A times Partner B while Partner B puts the puzzle together. Partner A records the time on a piece of paper. Then Partner B times Partner A and records his or her time. Each partner receives five chances to solve the puzzle. After each trial, the pieces should be mixed or shuffled together.
5. Upon completion of the game (five trials per partner), calculate the average time it took your partner to solve the puzzle. *Hint:* Add the five scores and divide the total by five. The player with the lowest average time wins the game.



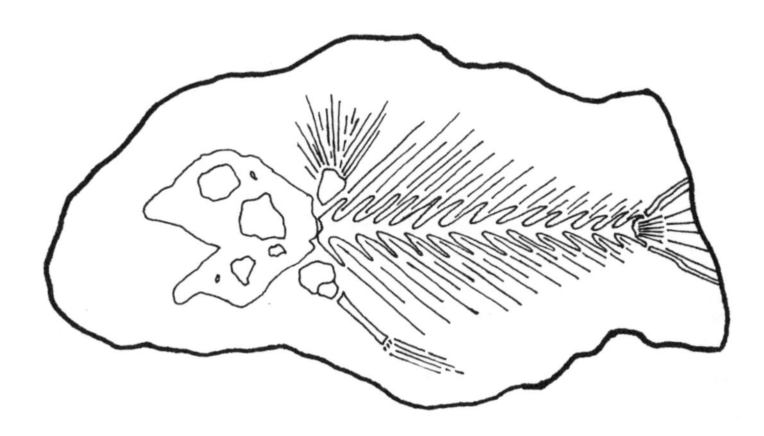

## Section 7: Fossils
## TEACHER'S GUIDE AND ANSWER KEY

### 7-1 What Is a Fossil?

***Part 1***

Statements 2, 3, 5, 6, 8, and 9 are true. Definition: A fossil is preserved evidence of past life.

***Part 2***

False statements 1, 4, 7, and 10 can be changed to true statements as follows:

1. *Not* all organisms die and become fossils.
4. Organisms with *hard* parts have the best chance of becoming fossils.
7. Tar pools trapped and preserved animals that lived during the *Pleistocene* era.
10. The process of metamorphism *can* distort or totally destroy fossil evidence.

***Part 3***

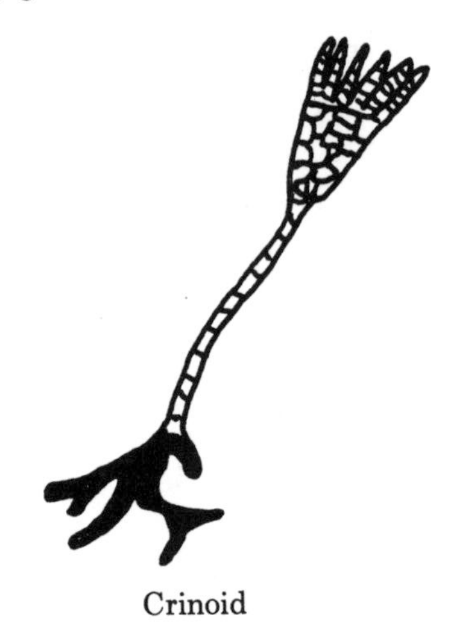

Crinoid

Plant leaf

Trilobite

Ammonite

## 7-2 Fossil Term Roundup

### *Part 1*

1. imprint; 2. scavenger; 3. carbonization; 4. amber; 5. sediment; 6. rock; 7. plant; 8. mold; 9. cast; 10. Mesozoic

### *Part 2*

jellyfish, snail, clam, worm, fern, coral, sponge, starfish, fly, cycad, fish, flower

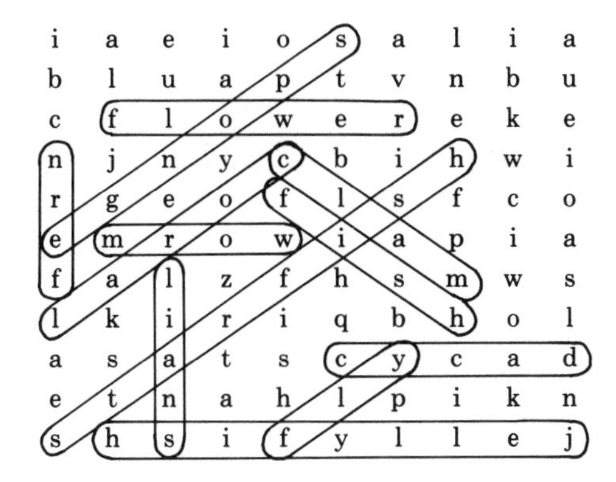


## 7-3 Who Was That Dinosaur?

### *Part 1*

1. c; 2. d; 3. a; 4. X; 5. e; 6. b; 7. X

### *Part 2*

1. b; 2. c; 3. a; 4. d; 5. e; 6. f

## 7-4 Fossil Clues

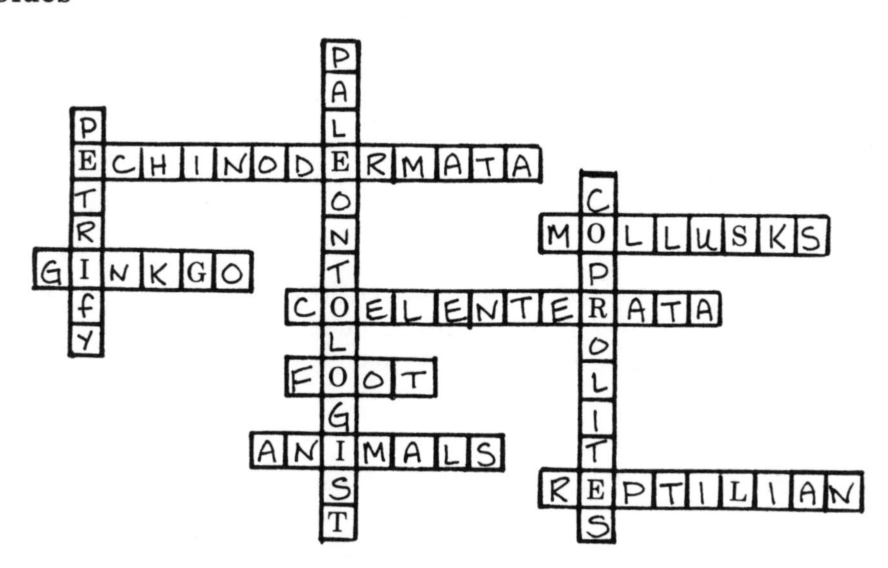


1. paleontologist; 2. petrify; 3. foot; 4. coprolites; 5. coelenterata; 6. echinodermata; 7. animals; 8. mollusks; 9. reptilian; 10. ginkgo

## 7-5 A Survey of Bones

If possible, let students examine a human skeleton prior to doing the activity. Students will see firsthand how hard and soft structures combine to form the human shape.

### *Part 1*

These are possible answers: *Hardest*—skull, femur, humerus, pelvic region, backbone, tibia; *Hard to soft*—ribs, wrist bones, temporal region of the skull, sternum; *Softest*—ligament, cartilage, tendon.

### *Part 2*

1. Skull fragments, teeth, jaw fragments; any thick, strong parts of a bone
2. They are made of a hard substance; proper conditions for fossilization
3. Soft rib cartilage decomposes.
4. To replace lost cartilage
5. Rib cage—plastic cartilage; screws and bolts to connect bones at the joint
6. Quick burial in undisturbed material; for example, tree sap hardening around an insect
7. Extreme heat and pressure; remains destroyed soon after death
8. Answers will vary.
9. How you might have looked, where you lived, something about your lifestyle, and so on
10. If the remains revealed apparent weaknesses in how a person lived, contemporary humans would know what problems to avoid.
11. Perhaps they were buried rapidly and left undisturbed.

## 7-6 Plaster Fossil Model

Have enough illustrations on hand for students to examine. Some, of course, will take longer than others to choose an organism. Tell students not to choose a detailed subject unless they have the patience and interest to complete the project.

Remind students to pour water slowly into the mixture while stirring. If the mixture hardens or doesn't set well, have them start over with fresh material.

## 7-7 Chalk Stick Fossil Model

Tell students to take their time and have patience. Brittle chalk tests the scraping ability of everyone.

Showcase the models for everyone to see. Many students appreciate it if their products are shown to other students, teachers, and parents.

1. Sedimentary rock
2. Great quantities of organisms died and became preserved at the same time.
3. School chalk is a manufactured gypsum product. Chalk, a sedimentary rock, is compact limestone composed largely of the shells of minute, marine organisms.

## 7-8 Wax Imprints

Remind students to keep their work area clean and to return materials to their respective places.

1. The material isn't fine enough to preserve the detail of an organism.
2. Poor conditions for fossilization; no organisms living in the area
3. Probably fig; it contains a tougher outside covering than the other organisms.

## 7-9 Molds and Casts

Tell students not to boil the wax. They need only heat the wax until it melts.

Students easily confuse mold and cast. You can simplify these terms by doing the following demonstration. Hold up a beaker half filled with dirt. Tell students the beaker represents a mold or cavity formerly occupied by an organism. The dirt acts as the filling material or cast.

### *Part 1*

1. The decomposition of soft body parts
2. It produces evidence of past life.
3. An organism that has decayed or dissolved
4. A shotput leaving a dent in the ground, a car tire rolling over soft dirt, and a person digging a hole in the ground
5. By providing preserved evidence of past life

### *Part 2*

1. Gradual buildup of Earth's sediment
2. The mold may also dissolve or wear away.
3. The external and/or internal structures of an organism
4. Iron pyrite, silica, dissolved mineral matter in water

## 7-10 Snail Cast

Collect dead garden snails in advance and let them dry out. A dried organism is easier to work with and less offensive to have around than a smelly, decaying snail.

The final products make excellent model fossil casts to exhibit. Many students accept the challenge and try to create the "perfect" cast.

1. Organism
2. Duplicating substance; i.e., filling material
3. Minerals, limestone, sandstone, and other sediments
4. Where the snail lived, its size and shape, etc.
5. Essentially the same things as listed in Question #4

## 7-11 Pencil Print

1. Sedimentary rock
2. Organism
3. Pressure exerted on the organism
4. A thin, carbonized film

## 7-12 Forever Amber, Part 1

If you feel this activity might be too messy for your classroom, have the students do it at home. Everything they need is readily available, and they can take their time completing the activity.

1. Smell, taste, curiosity, or simply by accident
2. By quick burial, no time for decomposing agents to break down tissue. Also, the resin covering kept air away from organisms.
3. Insects and tar animals were both trapped, covered, and preserved in a sticky substance. They differ in that insect remains show more detail.
4. More detailed structure to examine

## 7-13 Forever Amber, Part 2

Again, this activity may be completed at home. Home activities allow the more ambitious student or the student needing makeup work to earn credit at his or her leisure.

1. The simulation or model version calls for dead insects, not live organisms; mucilage is substituted for tree resin in the model version.
2. Both show how insects can be preserved in a sticky, resin-like substance.
3. Most likely yes. The mucilage keeps air and decay-causing bacteria from destroying the insects. If the mucilage dries and cracks, the exposed insects will eventually dry out and wither away.

## 7-14 "Reconstruct," a Fossil Game

This activity allows students to challenge their abilities to put a puzzle together. The pieces represent the end products of a fossil distorted by the metamorphic process.

## Section 7: Fossils
## MINI-ACTIVITIES

Here are 15 mini-activities, one to five minutes in length, for students to do at the beginning or end of the period.

1. Unscramble the words that complete the following sentence: Hard parts of (alnisma) ________________ are often (vedrrpese) ________________ in an (erdutneal) ________________ condition. (Answers: animals, preserved, unaltered)

2. Fill in the chart by writing the names of plants and animals that could become fossilized. Their names must start with the letter listed in the chart.

| Organism | F | S | P |
|---|---|---|---|
| Plant | | | |
| Animal | | | |

(**Answers:** *Plant*—(F) flower, fern, fig; (S) sycamore, spruce, sequoia; (P) palm, pine, plum. *Animal*—(F) frog, fish, fly; (S) sponge, spider, shellfish; (P) parasite, plesiosaur, pteranodon)

3. Complete the puzzle.

**Across**

1. Soft-bodied, slow-moving creature
3. A pattern or design
4. Sometimes wood becomes like this

**Down**

1. A mixture of dirt or mud
2. The beginning of a new geological system or formation

(**Answers:** *Across*—1. snail; 3. imprint; 4. petrified; *Down*—1. sediment; 2. era

4. Unscramble the following names of reptiles that roamed the Earth during the Mesozoic era.

suutrsnoboar sugusteosar oerspulsia

(**Answers:** brontosaurus, stegosaurus, plesiosaur)

5. See how many words you can get from the word *mesozoic.*
(**Answers:** so, ooze, me, is, some, zoo, moose, and so on)

6. Create a silly book title and author from fossil-related terms. For example: *From Snail to Mud* by Ima Mold

7. Complete the following analogies:

   Cenozoic is to era as cretaceous is to ________________.

   Whale is to mammal as dinosaur is to ________________.

   Limestone is to sedimentary as lava is to ________________.

   Insect is to amber as saber-tooth tiger is to ________________. (two words)

   (**Answers:** period, reptile, igneous, tar pits)

8. Use words related to fossil study to rhyme with the listed terms below:

   residence ____________________ petition ____________________

   detains ____________________ mystery ____________________

   deserve ____________________ display ____________________

   (**Answers:** evidence, remains, preserve, condition, geology, decay)

9. This is all that's left of a fossil. If you put the pieces together, they will spell out the name of a once-living organism.

   (**Answer:** fish)

10. This is all that's left of a plant leaf fossil. If you put the pieces together, they will spell out the name of a once-living organism.

    (**Answer:** maple)

11. Sketch an ancient reptile with the following description:
    - A tail one-half the length of its body
    - Strong, muscular hind legs
    - Front and hind legs adapted for climbing trees
    - A long, pointed head with sharp teeth

12. Identify the fossils based on the following bits of evidence:
    - spines, sea, tube feet, rays, rocks
    - large, high hump, horns (spread more than 6 feet), furry
    - wooly, large tusks, lived during Ice Age, elephant-like
    - tentacles part of foot, mantle, sea-dwelling, coiled shell

    (**Answers:** starfish, bison, mammoth, nautilus)

13. A 6-foot fish had the fossilized bones of a smaller fish in its stomach area. What three things about the past life of these fish do the remains reveal?

    (**Answers:** (1) Fish lived during this time; (2) Larger fish ate smaller fish; (3) Both fish died at about the same time.)

14. An organism was found fossilized in amber. Unscramble the code and discover the organism's identity:

    *One letter rhymes with day; another letter's sound is the opposite of "out"; one letter rhymes with "key."*

    Now put the letters in order.

    (**Answer:** ant)

15. Complete the last line of the limerick. The last line should have no more than nine syllables and rhyme with *Doris:*

    *There once was a scientist Doris,*
    *Who dug up a great brontosaurus;*
    *As she picked through the soil,*
    *And wrapped bones in foil,*

    ______________________________.

Section 8

# GEOLOGIC TIME SCALE

## OUTLINE


8-1 Geologic Time Scale Vocabulary Puzzle
8-2 Geologic Time Scale Game
8-3 How Much Do You Know About the Geologic Timetable?
8-4 Examining the Geologic Timetable
8-5 Rock Music Timetable
8-6 Paleozoic Life Scramble
8-7 Mesozoic Life Scramble
8-8 Cenozoic Era Word Search
8-9 Four Sides: A Geologic Timetable Game
8-10 Life: Past and Present


## MATERIAL

The following laboratory materials are needed for Section 8:

- earth science textbooks or geologic timetables
- 8½″ by 11″ paper
- paper cement
- teacher-prepared 9″ by 4″ envelopes

Name ______________________________ Date ____________________

# 8-1 Geologic Time Scale Vocabulary Puzzle

Complete the puzzle using words related to the geologic time scale.

## Across

3. A method for determining the age of a fossil
7. The Earth's geologic history divided into units of time
8. A major event in the Earth's history usually associated with mountain building
9. Distinct layers of sedimentary rocks
11. A stage in a geological period; Pliocene, for example
13. The Mississippian and Pennsylvanian periods grouped together
14. A unit of geologic time; Cretaceous, for example
15. The intervals between geological revolutions which mark great changes both in the rock record and the life record
17. The age of reptiles
20. A scientist who studies ancient life

## Down

1. This era makes up the majority of geologic time
2. The age of mammals
4. A term that refers to the type of life that dominated a time span; for instance, the time when reptiles dominated the Earth
5. The geological time period referring to chalk formation
6. A period of time in the Paleozoic Era that begins with "D"
10. The epoch when apes appeared
12. During this era the land was populated with plants, the first forests appeared, and coal swamps developed
16. Preserved evidence of past geological life
18. Trilobites became extinct during this period
19. The ruling group of reptiles during the Mesozoic Era

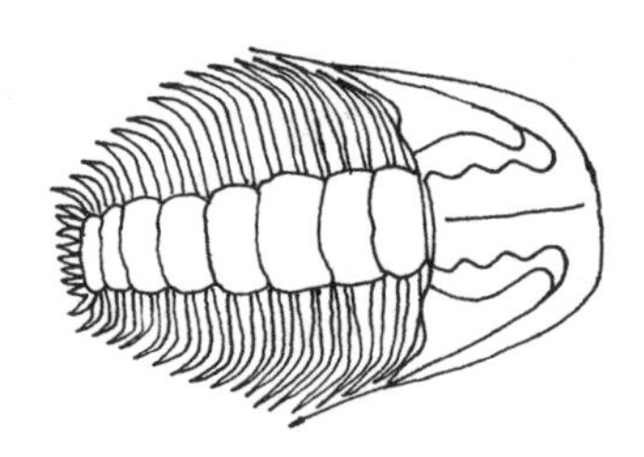

1
2
3
4
5
6
7
8
9
10
11
12
13
14
15
16
17
18
19
20

Name ______________________ Date ______________________

# 8-2 Geologic Time Scale Game

On the lines below, write the word that best fits the description on the left. When you are finished, the boxed letters will answer the question: What law did James Hutton, a Scottish geologist, believe helped explain certain geological events?

1. The animals characteristic of a geological period — _ _ [ ] _ _
2. A period of time in the Cenozoic Era — _ _ _ _ [ ] _
3. A type of fossil that indicates the geologic age of the rocks in which it is found — [ ] _ _ _ _
4. The plants characteristic of a geological period — [ ] _ _ _ _
5. A group of reptiles that lived during the Mesozoic Era — _ _ _ [ ] _ _ _ _ _
6. Rocks formed during this time make up a system. — _ _ [ ] _ _ _
7. A period of time that preceded the Ordovician Period — _ _ [ ] _ _ _ _ _
8. The Pleistocene is called the ______________ Age. — [ ] _ _
9. A period in the Cenozoic Era — _ _ _ [ ] _ _ _ _
10. Tiny sea animals that live in colonies; abundant in the Permian Period — _ _ _ [ ] _ _
11. The Miocene Epoch occurred during this period. — _ _ [ ] _ _ _ _ _
12. The first land animals (spiders, scorpions) appeared during this period. — _ [ ] _ _ _ _ _ _
13. The longest segment of geologic time — _ _ [ ]
14. Today we live in the ______________ Era. — _ _ [ ] _ _ _ _ _
15. A division of an era — _ _ _ [ ] _ _
16. Periods are divided into these smaller time units. — _ _ _ _ _ [ ]
17. The Cenozoic Era is known as the Age of ______________. — _ _ _ [ ] _ _ _

Name ______________________ Date ______________________

# 8-3 How Much Do You Know About the Geologic Timetable?

See how well you do on this 20-question true-false quiz. Don't worry about failing the test. There are no grades, only a series of short statements explaining your test score. Place *T* for *True* and *F* for *False* in the space to the left of the number.

___ 1. The dates and information on the geologic timetable are fixed and will never be changed.

___ 2. The fossil record determines the beginning and end of an era.

___ 3. Dinosaurs died out in the Triassic Period.

___ 4. The Silurian is the oldest period of time in the Paleozoic Era.

___ 5. Mesozoic means "Middle Life."

___ 6. Epochs are longer and more distinct units of time than periods.

___ 7. The oldest era, Archeozoic, contains rocks loaded with fossils of land animals.

___ 8. The Ice Age occurred during the Pliocene Epoch.

___ 9. Cenozoic means "Recent Life."

___ 10. The Paleozoic Era is divided into six periods.

___ 11. The earliest known records of animal life are marine invertebrates.

___ 12. The transition from marine animals to land animals occurred during the Silurian Period.

___ 13. The Mesozoic Era is divided into two periods: Triassic and Tertiary.

___ 14. Evidence of early humans appeared during the Ice Age.

___ 15. It is possible to establish some points on the geologic time scale in terms of years by studying the rate of decay of radioactive minerals contained in some rocks.

___ 16. The term *cretaceous* refers to chalk.

___ 17. The Mesozoic Era lasted approximately 50 million years.

___ 18. The Cenozoic Era is known as the "Age of Reptiles."

___ 19. The Paleozoic Era is known as the "Age of Invertebrates."

___ 20. The Mesozoic Era is known as the "Age of Fish."

## SCORING

| Number of Items Correct | Observation |
|---|---|
| 20–18 | You're a beginning geologist; start packing your bags. |
| 17–14 | You know what time it is; you're standing on solid rock. |
| 13–10 | You're hanging in there; your parents would be proud. |
| 9–6 | Big deal. A rock is a rock, right? |
| 5 or less | Grab your spikes and glove and head for the baseball park. Your team needs a shortstop. |

Name ______________________ Date ______________________

# 8-4 Examining the Geologic Timetable

Refer to a geologic timetable in an earth science textbook to answer the following questions:

1. Many plants and invertebrates shared the environment with mammals during the Cenozoic Era. Why, then, is the Cenozoic Era known as the Age of Mammals, not the Age of Plants or Age of Invertebrates?

   ______________________

   ______________________

2. What evidence is there that birds may have evolved from flying reptiles? ______________________

   ______________________

   ______________________

3. Why do you think the Mesozoic Era is called "Middle Life"? ______________________

   ______________________

   ______________________

4. Why isn't the Devonian Period known as the Age of Reptiles or Age of Corals instead of the Age of Fish?

   ______________________

   ______________________

5. Why do you think scientists aren't exactly sure how long a period or an epoch lasted?

   ______________________

   ______________________

6. When scientists are unsure of time length, how do they generally label an amount of time?

   ______________________

7. Write three statements regarding the Paleozoic Era. (Example: The Paleozoic Era is divided into seven periods.)

   - ______________________
   - ______________________
   - ______________________

8. When did the ammonites die out? ______

   What do you think may have caused the ammonites to become extinct? ______

   ______

   ______

9. If you wanted to know more about the Proterozoic Era, what two questions might you ask?

   - ______
   - ______

10. If you wanted to know more about the Mesozoic Era, what two questions might you ask?

    - ______
    - ______

11. What do the Pliocene and Miocene epochs have in common? ______

    ______

    ______

12. What do the Mississippian and Pennsylvanian periods have in common? ______

    ______

    ______

13. In what two ways are the Triassic and Cretaceous periods alike? ______

    ______

    ______

14. In what two ways are the Triassic and Cretaceous periods different? ______

    ______

    ______

15. According to the information on the geologic timetable, how long have humans been on Earth? ______

Name ______________________ Date ______________________

# 8-5 Rock Music Timetable

Scientists have constructed a geologic timetable from various life forms and physical events—mountain-building processes—since the Archeozoic Era. Now it's your turn to construct a timetable based on the information provided in the chart. The chart lists information about rock music from 1955 to the present.

The guidelines are simple. Use dates (years) for *eras,* musical categories for *periods,* and groups/individual artists for *epochs.*

On your timetable, indicate about when a certain event started (year), how long it lasted (how many years), and a feature or two to describe the event. Add anything else you feel needs to be included on the timetable.

Refer to a geologic timetable to help you construct your Rock Music Timetable. You'll need the following materials: ruler, pencil or pen, geologic timetable, and paper.

Answer the following questions when you complete your timetable:

1. What are three major problems with constructing a timetable?
   - ______________________
   - ______________________
   - ______________________
2. What are two weaknesses of the Information Chart?
   - ______________________
   - ______________________
3. Would there be more than one version of the Rock Music Timetable? Why or why not?

   ______________________
4. In what three ways is the Rock Music Timetable like a geologic timetable?
   - ______________________
   - ______________________
   - ______________________
5. How is the Rock Music Timetable different from a geologic timetable?

   ______________________

   ______________________

   ______________________

   ______________________

   ______________________

# Information Chart

| Year(s) | Musical Categories | Groups/Artists | Features |
|---|---|---|---|
| 1955 | rhythm and blues; rock and roll | Chuck Berry<br>"Fats" Domino<br>The Midnighters | appeal to teenagers |
| 1955–1959 | rock and roll | Bill Haley & the Comets<br>Little Richard<br>Elvis Presley | appeal to teenagers |
| 1960s | rock became socially radical, drug-alluding music. | The Beatles<br>Rolling Stones<br>Other English groups | English influence; electrical instrumentation |
| 1960s | movement toward acid rock; rock and roll | Jefferson Airplane<br>Bob Dylan<br>Jimi Hendrix | electrical amplification; additional instruments; light shows, etc. |
| 1970s | folk rock, country rock, jazz-rock fusion; rock and roll; instrumentation, synthesizers, etc. | Alice Cooper<br>Carole King<br>Led Zeppelin | hard-rock styles; amplified rhythms |
| middle 1970s | soft rock | Hall and Oates<br>Boz Skaggs | Soft, easy listening to a smooth beat |
| late 1970s | rock and roll, disco | Bee Gees<br>Donna Summer<br>Boston<br>Rod Stewart | uncomplicated, rhythmic music |
| 1980s and 1990s | rock and roll, punk rock, heavy metal | Whitesnake<br>Plasmatics<br>Aerosmith<br>Van Halen<br>Journey | loud, hard rock; extremes in costume and staging |

Name ______________________________ Date ______________________

# 8-6 Paleozoic Life Scramble

The boxes below contain the scrambled names of organisms characteristic of each period in the Paleozoic Era. Unscramble the letters and write the names in the space provided. Above each box write the name of the period. The periods, from oldest to youngest, are as follows: Cambrian, Ordovician, Silurian, Devonian, Carboniferous (Mississippian and Pennsylvanian periods), and Permian.

1. ______________________________

s s n c o o p r i

______________

s m i e d l l e p i

______________

First d a n l l s p t a n

______ ________

s c l o r a

________

2. ______________________________

Cone-bearing l n a p s t

________

s l i a t s r o h e

______________

First p e r s e l t i

__________

s c i t n s e

__________

3. ______________________________

Primitive s h e f s i

________

s n a d u i o t l i

______________

s t e r i t i l b o

______________

4. ______________________________

b o l s e t i i r t

______________

y l e j l h i f s

____________

b h c o i a r d s o p

________________

s s l i n a

________

5. ______________________________

First s m n p a h a i b i

________________

h i f s

_____

Wingless i s s c t e n

__________

Large l d n a n l s t a p

______ ________

6. ______________________________

e d s e plants

_____

mammal-like l e s p r t i e

___________

s l o c r a

________

The Paleozoic Era is known as the Age of ______________________________.

Name ______________________ Date ______________

# 8-7 Mesozoic Life Scramble

The boxes below contain the scrambled names of organisms characteristic of each period in the Mesozoic Era. Unscramble the letters and write the names in the space provided. Above each box write the name of the period. The periods, from oldest to youngest, are as follows: Triassic, Jurassic, and Cretaceous.

1. ______________________

| | |
|---|---|
| Giant d c a y c s ________ | s e l i t e r p thrived ________ |
| First i r d s b ________ | a s d m i m o n o ________ |
| i s f o n e c r ________ | Primitive l m m s a m a ________ |

2. ______________________

| | | |
|---|---|---|
| g f l n i y reptiles ________ | c m a l ________ | r t s y e o ________ |
| s b t o e l r ________ | a s n m g r i s e p o ________ | s d r i b ________ |
| a l s m m m a ________ | Marine s l p r e i e t ________ | r o n i d u s a s ________ |

3. ______________________

| | | |
|---|---|---|
| Giant i r u a s s o n d ________ | First d r i s b o s r i d i c n ________ ________ | |
| Abundant c s c a d y ________ | m n d s a m o i o ________ | s b e l e m d i o n ________ |
| Numerous s n o c r e f i ________ | Small m m s l a m a ________ | |

The Mesozoic Era is known as the Age of ______________________.

Name ______________________ Date ______________________

# 8-8 Cenozoic Era Word Search

Locate and circle the 20 organisms commonly found in the Cenozoic Era. The names are listed below the puzzle. They may be found backward, forward, vertically, horizontally, and diagonally.

| | | | | | | | | | | | |
|---|---|---|---|---|---|---|---|---|---|---|---|
| c | o | n | i | f | e | r | s | a | e | a | b |
| i | e | a | s | h | t | o | m | m | a | m | i |
| s | f | m | a | b | e | a | r | s | u | p | r |
| u | s | t | i | u | r | f | r | i | f | h | d |
| p | c | f | l | o | w | e | r | i | d | i | s |
| p | a | i | e | a | t | e | s | c | e | b | s |
| i | w | g | r | s | h | h | j | a | g | i | e |
| h | h | s | y | t | e | a | p | m | e | a | s |
| o | a | o | a | s | r | u | m | e | l | n | s |
| s | l | g | r | e | p | t | i | l | e | s | a |
| e | e | o | h | i | p | p | u | s | i | e | r |
| m | s | n | o | d | o | l | i | m | s | a | g |

| | | |
|---|---|---|
| amphibians | figs | megatherium |
| bears | fishes | Mesohippus |
| birds | fruits | oysters |
| camels | grasses | reptiles |
| conifers | lemurs | smilodon |
| dire wolf | mammoths | whales |
| Eohippus | man | |

Name ______________________________ Date ______________________

# 8-9 Four Sides: A Geologic Timetable Game

In each puzzle below, one term is given. You must provide the remaining three terms. When all four terms are completed, they join together to form a square or rectangle. The answers may be written forward, backward, horizontally, and vertically. If you complete all four puzzles in 20 minutes or less, you'll be crowned a champion geologist!

1.

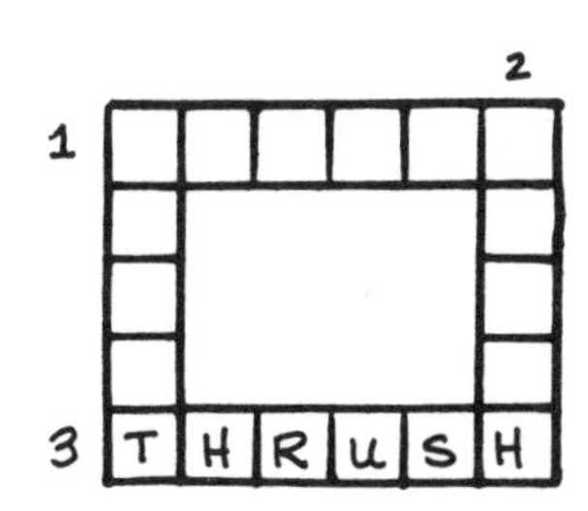


### Across

1. An epoch in the Cenozoic Era
3. A songbird with long wings and a long tail

### Down

1. To live
2. A solid-hoofed Cenozoic animal.

2.

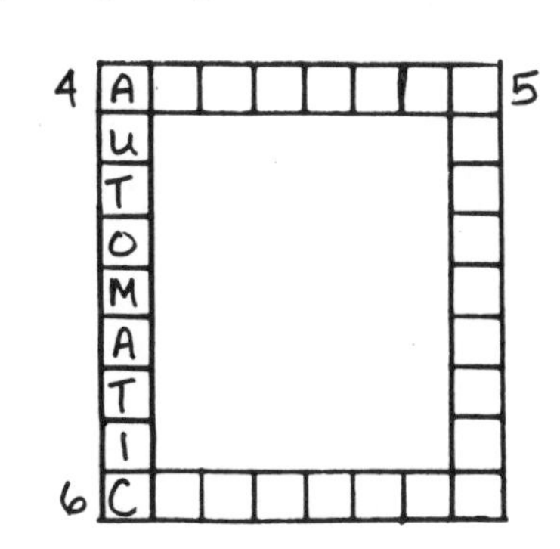


### Across

4. Dominant invertebrate of the Mesozoic Era
6. A period in the Mesozoic Era

### Down

4. The power of self-action
5. A Paleozoic crab-like marine animal

3.

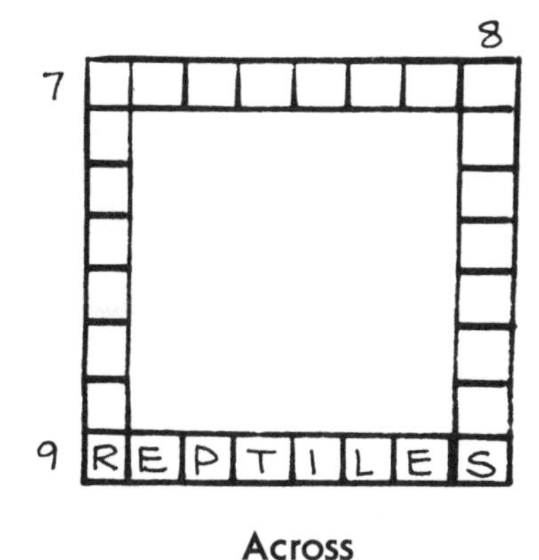


### Across

7. A period in the Paleozoic Era
9. A cold-blooded vertebrate which creeps or crawls on its belly

### Down

7. A "terrible lizard"
8. When land animals first appeared on Earth

4.

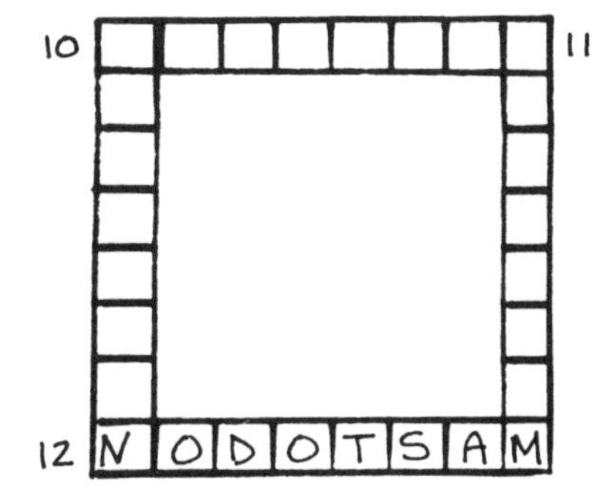


### Across

10. "Age of Mammals"
12. Extinct elephant-like mammal

### Down

10. Oldest period of the Paleozoic Era
11. "Age of Reptiles"

Name ________________________________ Date ________________________

# 8-10 Life: Past and Present

**Introduction:** Numerous living forms first appeared 600 million years ago, at the beginning of the Paleozoic Era. If life existed before this time, no fossil evidence has been found to support its existence. Since the Paleozoic Era, living organisms have proliferated on Earth.

**Objective:** To demonstrate the diversity of life throughout geologic history.

**Materials:** Teacher-prepared 9″ by 4″ envelopes, earth science textbook or geologic timetable, paper cement, and 8½″ by 11″ paper.

**Procedure:** Do the following:

1. Ask your teacher for a prepared envelope.
2. Open the envelope. Empty the contents on a desk or tabletop. Lay the four colored strips side by side. The brown strip represents the Cenozoic Era, the green strip represents the Mesozoic Era, the blue strip represents the Paleozoic Era, and the red strip represents the Proterozoic Era.
3. Refer to the Fossil Distribution Chart for information to do the following:
   a. Write the letter symbols of organisms on the colored paper which matches the corresponding geologic era(s).
   b. Paste the strips in the correct order (according to geologic history) on an 8½″ by 11″ piece of paper. Place them in a horizontal pattern, one-half inch apart.
   c. Attach the Fossil Distribution Chart to the assignment sheet.

Answer these questions:

1. What do you observe regarding the diversity of life as it appears on the assignment sheet?

______________________________________________________________

______________________________________________________________

2. What, in your opinion, are three weaknesses of the Fossil Distribution Chart?

- ______________________________________________________________
- ______________________________________________________________
- ______________________________________________________________

3. What, in your opinion, are three strengths of the Fossil Distribution Chart?

- ______________________________________________________________
- ______________________________________________________________
- ______________________________________________________________

4. How can learning about the diversity of life help you once you leave school?

____________________________________________

____________________________________________

## Fossil Distribution Chart

| Symbol | Organism | Time Distribution |
|---|---|---|
| Cr | Crinoids | Paleozoic to Cenozoic |
| R | Reptiles | Paleozoic to Cenozoic |
| Amp | Amphibians | Paleozoic to Cenozoic |
| B | Birds | Mesozoic to Cenozoic |
| WM | Woolly mammoths | Cenozoic |
| T | Trilobites | Archeozoic to Paleozoic |
| Br | Brachiopods | Proterozoic to Cenozoic |
| STC | Saber-tooth cats | Cenozoic |
| Ich | Ichthyosaurs | Mesozoic |
| Bc | Bacteria | Archeozoic to Cenozoic |
| A | Ammonites | Mesozoic |
| F | Ferns | Paleozoic to Cenozoic |
| Fh | Fishes | Paleozoic to Cenozoic |
| In | Insects | Paleozoic to Cenozoic |

# Section 8: Geologic Time Scale
# TEACHER'S GUIDE AND ANSWER KEY

## 8-1 Geologic Time Scale Vocabulary Puzzle

*Across:* 3. radioactive dating; 7. time scale; 8. revolution; 9. strata; 11. epoch; 13. carboniferous; 14. period; 15. era; 17. Mesozoic; 20. paleontologist; *Down:* 1. Precambrian; 2. Cenozoic; 4. age; 5. cretaceous; 6. Devonian; 10. Miocene; 12. Paleozoic; 16. fossil; 18. Permian; 19. dinosaurs

## 8-2 Geologic Time Scale Game

*Tertiary* appears twice in the puzzle (9 and 11). You may want to tell students that the same period of time in the Cenozoic Era appears twice in the puzzle.

1. fauna; 2. Eocene; 3. index; 4. flora; 5. dinosaurs; 6. period; 7. Cambrian; 8. ice; 9. Tertiary; 10. corals; 11. Tertiary; 12. Silurian; 13. era; 14. Cenozoic; 15. period; 16. epochs; 17. mammals.

**Answer to question:** UNIFORMITARIANISM

## 8-3 How Much Do You Know About the Geologic Timetable?

This is a fun quiz you can give students anytime. The items will test their knowledge and guessing ability. The scoring system is presented in a tongue-in-cheek manner so students won't feel anxious about earning a high score.

1. false, under constant revision; 2. true; 3. false, cretaceous; 4. false, Cambrian; 5. true; 6. false, shorter and less distinct; 7. false, almost no fossils; 8. false, Pleistocene; 9. true; 10. false, seven periods; 11. true; 12. true; 13. false, Jurassic; 14. true; 15. true; 16. true; 17. false, 125 million; 18. false, mammals; 19. true; 20. false, reptiles

## 8-4 Examining the Geologic Timetable

It may be easiest to give each student a copy of the geologic timetable. Ask students to answer the 16 questions based on the information provided on the timetable or on what they might infer from studying the timetable.

***Possible Answers:***

1. Mammals made an appearance during this time. Plants and invertebrates were already on Earth for millions of years.
2. Birds have structural similarities to the flying reptiles.
3. The Mesozoic Era appears near the middle of the timetable, midway between the Archeozoic Era and Cenozoic Era.
4. Numerous species of fish dominated this particular period of geologic history.
5. Answers will vary. Probably because the dating techniques are not 100% accurate.
6. Probable duration
7. Answers will vary.
8. The ammonites died out in the Mesozoic Era. Seas may have retreated, possibly adverse climatic changes, drastic changes in sea temperature, dwindling food supply, and so on

9. Answers will vary.
10. Answers will vary.
11. Horses were present in both epochs, considerable volcanic activity in various parts of the world, both epochs are part of the Cenozoic Era, both grouped under the Tertiary period, and so forth
12. Both periods are part of the Paleozoic Era, plants and amphibians flourished in both periods, both lasted about 30 million years, and so on
13. Both periods are part of the Mesozoic Era, dinosaurs lived during both periods of time, mammals lived during both periods of time, and so forth
14. The Cretaceous lasted almost twice as long as the Triassic, reptiles thrived during the Triassic and died out in the Cretaceous, and so on
15. About two million years

## 8-5 Rock Music Timetable

Students may have trouble getting started, so you might show them one or two examples of setting up a timetable. The finished products make excellent items to display around the classroom. *Note:* This activity will take more than one class period to complete.

***Possible Answers:***

1. a. Placing items in the correct time slots
   b. Disagreement with certain aspects of the Information Chart
   c. Wanting to include more information (in addition to the Information Chart) on the Rock Music Timetable
2. a. At best, it provides a minimal amount of information.
   b. It offers only one person's (the author's) version of the history of rock music.
3. Yes. The information allows for flexibility; i.e., one category may overlap another. Also, a student's bias enters the construction of a rock music timetable.
4. a. It groups dates, events, and features.
   b. It shows the changes and evolution of events over time.
   c. It breaks down events into smaller units of time or time slots.
5. It categorizes music, not geological events; the earliest date appears at the top of the chart, not the bottom; the music information is easier to obtain. It doesn't rely on geological dating procedures.

## 8-6 Paleozoic Life Scramble

1. *Silurian:* scorpions, millipedes, first land plants, corals
2. *Carboniferous:* cone-bearing plants, horsetails, first reptiles, insects
3. *Ordovician:* primitive fish, nautiloids, trilobites
4. *Cambrian:* trilobites, jellyfish, brachiopods, snails
5. *Devonian:* first amphibians, fish, wingless insects, large land plants
6. *Permian:* seed plants, mammal-like reptiles, corals

The Paleozoic Era is known as the Age of Invertebrates.

## 8-7 Mesozoic Life Scramble

***From Left to Right:***

- *Triassic:* giant cycads, reptiles thrived, first birds, ammonoids, conifers, and primitive mammals

- *Cretaceous:* flying reptiles, clam, oyster, lobster, birds, angiosperms, mammals, marine reptiles, and dinosaurs
- *Jurassic:* giant dinosaurs, first birds, crinoids, abundant cycads, ammonoids, belemnoids, numerous conifers, and small mammals

The Mesozoic Era is known as the Age of Reptiles.

## 8-8 Cenozoic Era Word Search

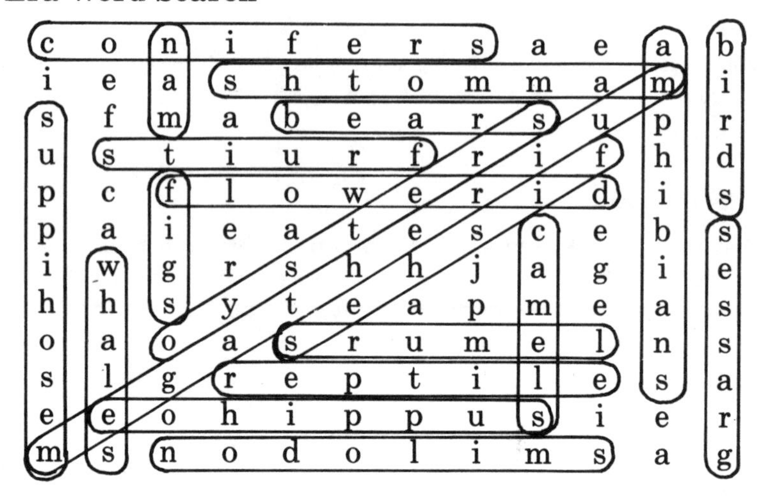


## 8-9 Four Sides: A Geologic Timetable Game

1. 2. 3. 4.

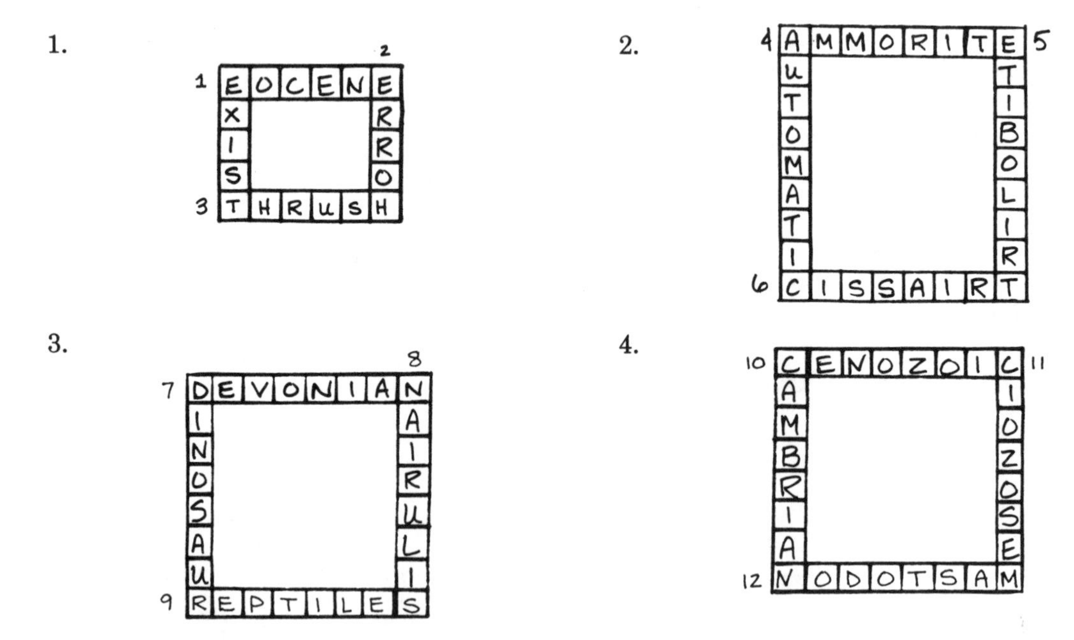


## 8-10 Life: Past and Present

Prepare an envelope for each student doing the activity. Place the following items in each envelope: four colored paper strips, 6″ by 2″, to represent four geologic eras—Cenozoic (brown), Mesozoic (green), Paleozoic (blue), and Proterozoic (red). Be sure to copy enough Fossil Distribution Charts for each student.

***Possible Answers:***

1. Organisms evolve or change into complex life forms.
2. a. It doesn't provide illustrations to show what some organisms looked like.
   b. It shows only a small percentage of organisms.
   c. It fails to provide any information regarding how or where an organism lived.
3. a. It's simple to read and understand.
   b. It highlights some of the more popular or well-known organisms.
   c. It condenses millions of years of life into a single page.
4. It can help you see how the strong survive and the weak perish. Also, like the old saying, it takes all kinds to make a world (ancient or modern). And most everyone—to some degree—depends on others for subsistence.

# Section 8: Geologic Time Scale
## MINI-ACTIVITIES

Listed below are 10 mini-activities, one to five minutes in length, for students to do at the beginning or end of the period.

1. A part of an ear and a bone located between the waist and shoulder can be spelled from the letters in trilobite. What are these structures?
   (**Answers:** rib and lobe)
2. **Riddle:** How did the mother dinosaur decorate the counter in her kitchen?
   (**Answer:** in rep*tile*)
3. **Riddle:** During the Paleozoic Era, when was a fish not a fish?
   (**Answer:** When it was a jellyfish.)
4. Match the periods below with the life characteristic of those times. Draw a line connecting the period with the life characteristic.

| | |
|---|---|
| Silurian | first angiosperms |
| Triassic | first amphibians |
| Devonian | first land plants |
| Jurassic | first mammals |

   (**Answers:** Silurian—first land plants; Triassic—first mammals; Devonian—first amphibians; Jurassic—first angiosperms)
5. **Riddle:** Where do most fossils spend the night?
   (**Answer:** in sedimentary beds)
6. Use the numbers, words, and letters to decode the names of the geological periods or epochs.
   - 3 + donkey-like + ek
     (**Answer:** Triassic)
   - p + lice + II + past tense of "saw"
     (**Answer:** Pleistocene)
   - sound of a happy cat + me + and (minus the d)
     (**Answer:** Permian)
   - sounds like "Oh, let go" + c + na
     (**Answer:** Oligocene)
7. If Alley Oop (fictional cartoon caveman) actually lived, during what geological time periods would he feel at home?
   (**Answer:** Triassic, Jurassic, or Cretaceous) If students ask why, tell them Alley Oop had a dinosaur for a pet.
8. **Riddle:** What Pleistocene animal got stuck in a tar pit but kept his sense of humor?
   (**Answer:** Smilodon, also known as the saber-tooth cat)

9. Notice that era names end in "zoic." Make up era names ending in "zoic" for the following descriptions:
   - 600 million years of listening and dancing to punk rock music
   (Answers will vary.)
   - 300 million years of communicating with hand gestures only
   (Answers will vary.)
   - 450 million years of living in the streets
   (Answers will vary.)
10. Use the one-word clue to identify the organism now living in the Cenozoic Era:

    cigarette (camel)
    radish (horse)
    cocktail (fruit)
    chowder (clam)
    Dumbo (elephant)
    ham (pig)

Section 9

# FORCES THAT SHAPE THE EARTH'S SURFACE

## OUTLINE


9-1 Restless Earth Term Puzzle
9-2 Erosion at Work
9-3 Weathering Processes
9-4 Forces Around Us, Part 1
9-5 Forces Around Us, Part 2
9-6 Glaciers on the Move
9-7 Sleuthing the Newspaper
9-8 Changing Earth Word Search
9-9 Diastrophic Events
9-10 Pancake Earth


## MATERIAL

The following laboratory materials are needed for Section 9:

- beakers (100 ml)
- hot plates
- magnifiers
- metric rulers
- pancake flour or Bisquick™
- Pyrex™ or metal trays
- teaspoons
- thermometers (F°)
- water

Name ______________________ Date ______________________

# 9-1 Restless Earth Term Puzzle

Complete the puzzle using terms related to the Earth and events responsible for shaping our planet.

### Across

2. A crack in the Earth's crust
4. The process of deformation of the Earth's crust, producing oceans, mountains, and ocean basins
7. The return to original shape after being stretched or twisted
9. The Earth's outer surface layer
11. An upfold in rock layers
14. A shaking or trembling of the Earth caused by sudden rock movement
15. Force or stress that tends to decrease the volume of a substance
16. A strong forward motion
17. The state of balance of the Earth's crust
18. Molten rock materials below the Earth's surface
20. Movement of heat by currents within the heated material; occurs only in liquids and gases

### Down

1. The movement of land in an upward direction
3. To distort from natural shape
5. A downfold of rock layers
6. An open fracture in a rock surface
8. A place where lava reaches the surface
10. Stress or strain on rocks
12. Breaks that occur in rocks
13. Results when two adjacent parts of a solid slide past one another parallel to the plane of contact
18. The layer of Earth located between the core and the crust
19. A bend in rock strata

## 9-1 Restless Earth Term Puzzle, continued

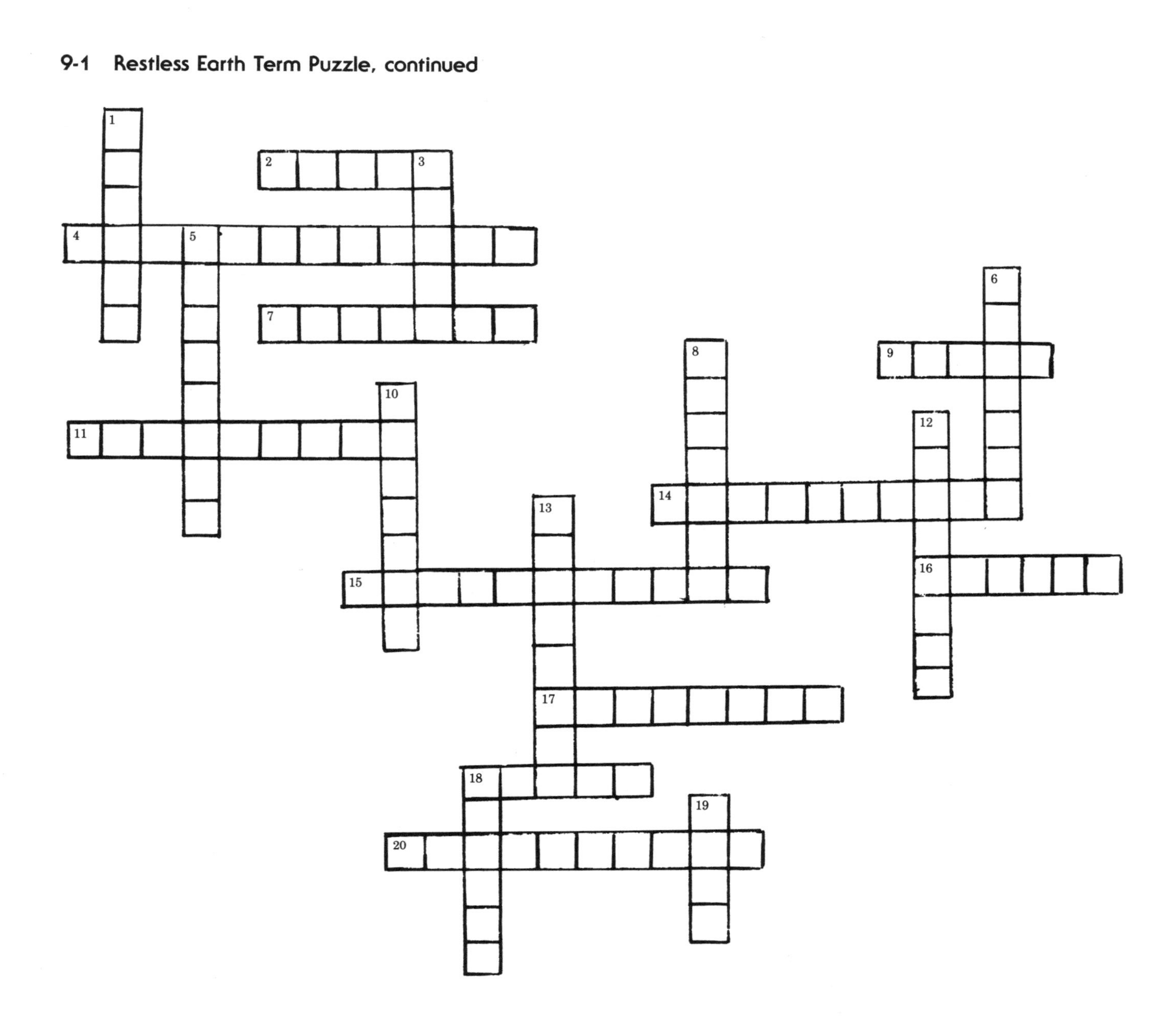

Name ______________________ Date ______________________

# 9-2 Erosion at Work

Erosion is the gradual wearing away of land by water, wind, and ice.

## PART 1

Find the answers to the hints listed and write them in the space provided. Then locate each term in the puzzle and shade it. The seven terms in the puzzle relate to erosional agents or the effects of erosion, and may be found vertically or horizontally. The shaded answers will answer this question: What do you call a deposit of clay, sand, and gravel left by a glacier?

______________________

1. Rock fragments or particles that collect at the base of a steep mountain

   ______________

2. Soil particles larger than clay particles, but smaller than sand grains

   ______________

3. A deposit at the mouth of a stream; triangular in shape ______________
4. Hills of loose sand heaped up by the wind ______________
5. Dry, flat lands worn down by erosional action ______________
6. An agent of erosion; air molecules in motion ______________
7. A combination of two different gases; an agent of erosion ______________

| | | | | | | | | | | | | | | | | |
|---|---|---|---|---|---|---|---|---|---|---|---|---|---|---|---|---|
| w | a | t | e | r | a | f | e | d | l | o | e | s | d | i | c | e |
| r | s | a | e | m | b | w | r | u | e | r | o | d | e | h | p | a |
| a | a | l | t | e | r | i | o | n | o | i | l | s | l | u | m | p |
| i | n | u | e | s | a | n | d | e | b | a | s | h | t | k | a | c |
| n | d | s | g | a | l | d | e | s | i | l | t | e | a | r | i | d |

## PART 2

Underline the word or phrase that correctly completes each of the following sentences:

1. A moving glacier uproots rocks and carries them along a forward path. This is called (abrasion, plucking, depositions).
2. The outside cracking or peeling of a rock is known as (exfoliation, sublimation, shearing).
3. Wind deposits or (barchan, fluvian, loess) are made up of silt and dust-size particles.

4. The deposit of gravel, sand, and boulders known as (compost, moraine, eolian) is evidence of a melting glacier.
5. Rocks in motion rub against each other. The friction created causes them to (aggregate, abrade, aerate).
6. Round depressions or (sinks, nodules, potholes) are made in the bed of a fast-moving stream.
7. A deposit of sediment or (anticline, alluvial fan, delta) forms when a mountain stream of high velocity enters a relatively flat valley.
8. A (rainstorm, glacier, tributary) may cause a huge flow of mud to move down a slope suddenly.
9. A (loess, barchan, windbreak) is a barrier that causes wind to lose speed.
10. The movement of (crevasses, glaciers, cirques) can erode and shape the land.



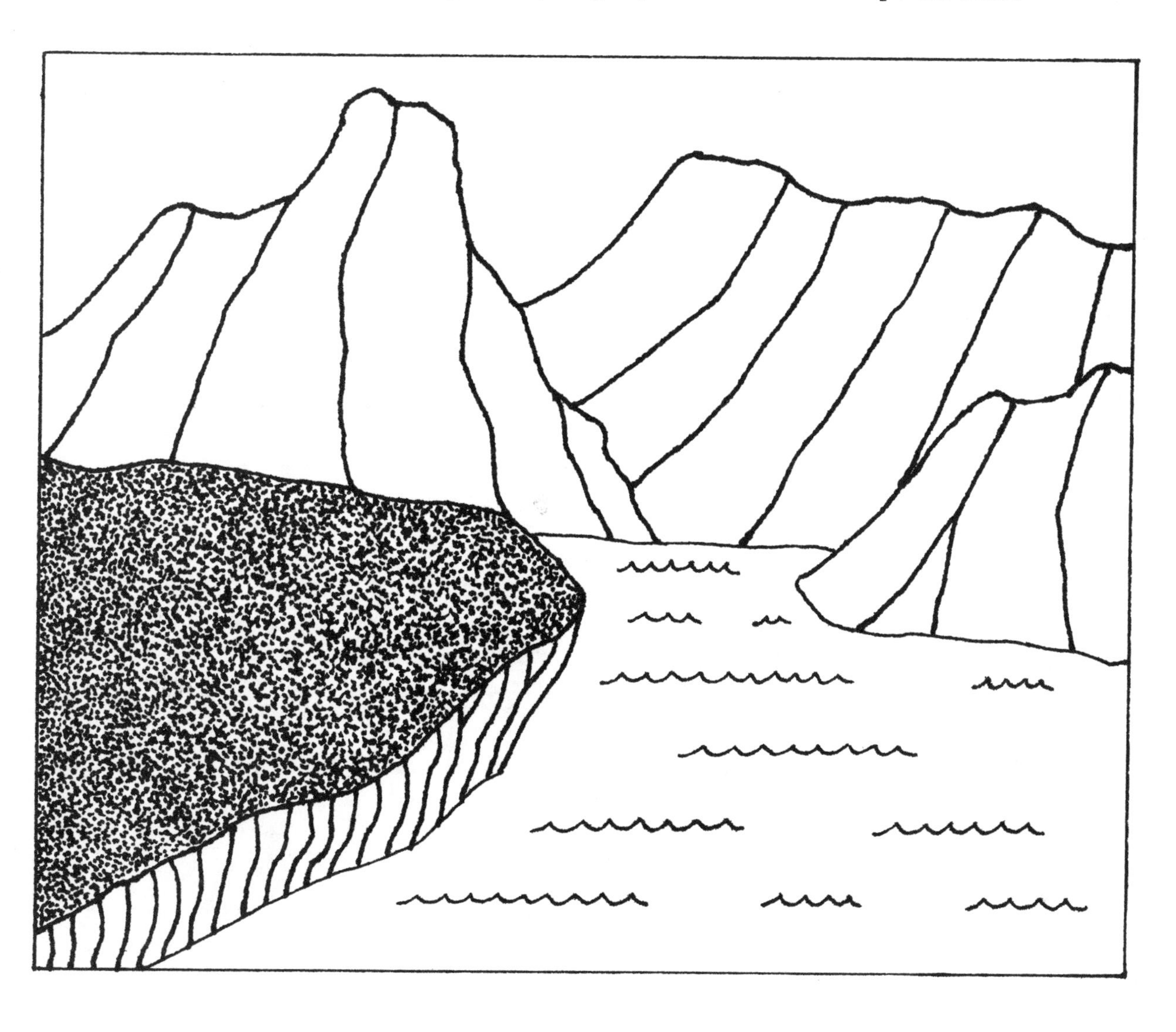

Name ______________________ Date ______________________

# 9-3 Weathering Processes

The natural process that breaks rock into smaller particles is called weathering. If rock breaks into smaller pieces without any change in the rock materials, physical weathering has occurred. Conversely, if some of the rock materials change, chemical weathering has occurred.

Four weathering processes are described below. Read each description. Then, in the space provided, sketch the process and label each item in the process.

1. Plant roots grow into the cracks in rocks. As the roots grow, they extend deeper into the rocks forcing the rocks to break apart.

This is an example of (physical, chemical) weathering. (Circle one.)

Explain your answer. ______________________

______________________

______________________

______________________

2. Water, like plant roots, fills cracks in a rock. Cool night temperatures cause the water to freeze. As you know, when water freezes, it expands. If there is freezing at night and thawing during the day, the rock may break apart.

This is an example of (physical, chemical) weathering. (Circle one.)

Explain your answer. ____________________________________________

____________________________________________

____________________________________________

____________________________________________

3. Running water in a stream is dissolving the minerals in Rock X. As the stream removes the minerals, Rock X becomes weak. Some of the remaining minerals absorb water, expand, and strain the rock. The weakened rock begins to break apart.

This is an example of (physical, chemical) change. (Circle one.)

Explain your answer. ____________________________________________

____________________________________________

____________________________________________

____________________________________________

4. The sun constantly heats a rock. Heat causes minerals within the rock to expand. As these minerals expand, the rock weakens and begins to crumble.

This is an example of (physical, chemical) change. (Circle one.)

Explain your answer. ____________________________________________

____________________________________________

____________________________________________

____________________________________________

Name ______________________ Date ______________________

# 9-4 Forces Around Us, Part 1

## PART 1

Use the hints in each statement to identify the force or the changes that occur due to the forces acting on the Earth. Use the boxed word list to select your answers. Be careful! Some words may be used only once; others may not be used at all.

**Events**

1. Mineral A changes into Mineral B. Mineral A, a hard, black object, decomposes into a reddish powder.

   Answer: ______________________

2. A slow-moving stream wears down its banks. Sediments are carried away by the stream.

   Answer: ______________________

3. As pressure decreases when overlying rock is removed by erosion, Rock X expands. The expansion produces cracks or breaks.

   Answer: ______________________

4. In mountainous and high plateau regions, a fast-moving river cuts rapidly into its bed. The river walls widen. What do these young rivers form?

   Answer: ______________________

5. Stresses build up in the Earth's crust. When the stress becomes too great, rocks break.

   Answer: ______________________

6. A riverbed widens and becomes more snake-like as it spreads out.

   Answer: ______________________

7. A river deposits sediments at its mouth. The accumulating sediments form a level, fan-shaped pattern.

   Answer: ______________________

8. This is a large mass of intrusive igneous rock of unknown depth.

   Answer: ______________________

| | | |
|---|---|---|
| anticline | erosion | physical weathering |
| batholith | geyser | syncline |
| chemical weathering | joints | uplift |
| delta | laccolith | V-shaped valley |
| dike | meander | volcano |
| earthquake | | |

## PART 2

Tell how each of the events previously listed help shape the surface of the Earth.

1. ______________________________
______________________________
______________________________

2. ______________________________
______________________________
______________________________

3. ______________________________
______________________________
______________________________

4. ______________________________
______________________________
______________________________

5. ______________________________
______________________________
______________________________

6. ______________________________
______________________________
______________________________

7. ______________________________
______________________________
______________________________

8. ______________________________
______________________________
______________________________

Name ______________________ Date ______________________

# 9-5 Forces Around Us, Part 2

In the space provided, sketch and label the main parts of each event described in Activity 9-4: Forces Around Us, Part 1. Then tell what economic effect, if any, each event has on a person's life. For example, a flood might ruin homes, buildings, and pollute the public water supply.

1.

Economic effect: ______________________

______________________

2.

Economic effect: ______________________

______________________

3.

Economic effect: ______________________

______________________

4.

Economic effect: ______________________________

______________________________

5.

Economic effect: ______________________________

______________________________

6.

Economic effect: ______________________________

______________________________

7.

Economic effect: ______________________________

______________________________

8.

Economic effect: ______________________________

______________________________

Name ______________________ Date ______________________

# 9-6 Glaciers on the Move

Slow-moving ice packs known as glaciers produce erosion by carving out valleys along their paths. A glacier forms when snow piles up deeper and deeper, and the lower snow layers are compressed into ice. The ice may begin to move down from a mountain slope in a slow-moving stream. The ice may be over 1,000 feet thick in places. Thus, a glacier is any large mass of ice on land capable of moving from one point to another.

## PART 1

What force acts on the ice and causes it to flow? (*Hint:* The flowing motion caused by the pressure of the ice combines with this force.) Find out by providing the correct answers to each description below. Fill in the blanks; then unscramble the circled letters to uncover the answer.

1. A long, slow-moving, wedge-shaped stream of ice (two words)

2. Deposit of earth materials left by a melting glacier

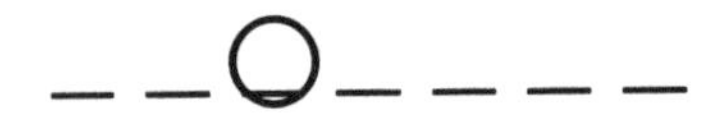

3. A fissure or chasm-like opening in the ice of a glacier

4. A steep-walled basin at the head of a glacial valley

5. Pertaining to or caused by ice masses

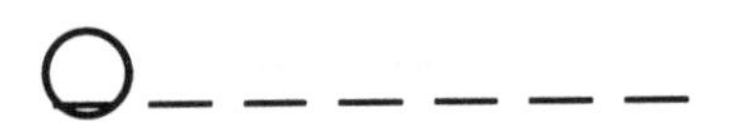

6. Depression left in glacial deposits when covered ice block melts

7. The granular compressed snow which forms glacial ice

The force that acts on the ice and causes it to flow is ______________________.

## PART 2

Match the term in the left-hand column with its description in the right-hand column. Draw a line connecting the term with the description.

| Term | Description |
|---|---|
| drumlin | circular hollow |
| horn | sharp ridge |
| cirque | parallel scratches |
| crevasse | glacial till |
| arête | rough ice |
| striations | sharp peak |
| moraine | steep basin |
| firn | glacial fissure |
| valley | alpine glacier |
| kettle | glacial hill |

## PART 3

Answer these questions:

1. How do glaciers change the shape of the Earth? ______________________________

_____________________________________________

_____________________________________________

_____________________________________________

2. In what ways are glaciers beneficial to humans? ______________________________

_____________________________________________

_____________________________________________

_____________________________________________

3. In what ways are glaciers harmful to humans? ______________________________

_____________________________________________

_____________________________________________

_____________________________________________

4. How can learning about glaciers increase your understanding of natural forces? ______

_____________________________________________

_____________________________________________

_____________________________________________

Name ______________________________ Date ______________________

# 9-7 Sleuthing the Newspaper

Be a newspaper sleuth. Carefully examine the newspaper for five or six days. Look for articles or short news items related to events responsible for shaping the Earth's surface—for example, hurricanes, tornadoes, or floods on the rampage; earthquakes ripping up the land; or anything people are doing to alter the Earth's crust. After you read each article, enter the information on the chart below. An example is given.

Name of the newspaper: ______________________________

Date(s) of publication: ______________________________

| Earth-Shaping Event | Geographic Location of Event | How Might This Event Change the Shape of the Earth's Surface? |
|---|---|---|
| flooding | Columbia River, WA | erode away parts of the land; widen the river; sediment buildup |
| | | |
| | | |
| | | |
| | | |
| | | |
| | | |
| | | |
| | | |
| | | |
| | | |
| | | |
| | | |

Name ______________________________ Date ____________________

# 9-8 Changing Earth Word Search

The nineteen terms listed below relate to forces or events that shape the Earth's surface. Seventeen of the terms may be, for the most part, attributed to nature, or humans may have a small part in making them happen. Two of the terms, without a doubt, are man-made. Use blue to *circle* the terms that occur naturally (ten of them), use green to circle the terms that are partially created by humans (seven of them), and use red to circle the terms that are totally caused by humans (two of them). Terms may be found forward, backward, vertically, horizontally, and diagonally.

g d c y e h f e d i l s d n a l
i l d r i g b m r o t s z o e a
a c a m h n e r o s i o n c a p
b w i c d i s c u b m m j o g n
f b e a i l r n o w s e k n e a
g t k v y e a r a i i t a s f h
a l a n d m r v h k n t q t i s
s b u r i s e p u j a u m r o t
x f q e w s o x r c c b e u q r
t f h l a r k i r e l p a c e e
l a t e t a v e i w o d o t m a
k c r s e e e r c p v j f i k m
w e a i r k o d a n r o t o e v
j i e a f o j e n a c g d n f n
d k g l a c i s e h o n i u t a
m a v a l a n c h e i t l s i e
c i r r l b i u a w p d u o l c

- avalanche
- construction
- creek
- diastrophism
- earthquake
- erosion
- glacier
- hurricane
- landslide
- river
- storm
- stream
- tornado
- tsunami
- volcanism
- war
- waterfall
- waves
- wind

Name ______________________ Date ______________________

# 9-9 Diastrophic Events

Diastrophism is the movement of the solid part of the Earth. If the Earth shows evidence of movement in an upward direction, we say uplifting has occurred. Conversely, if the crust has sunk or moved downward, we say subsidence has taken place. A horizontal movement of the crust is called thrust.

## PART 1

The dark rectangles below represent sections of the Earth's crust. The arrows represent force exerted on the crust from the buildup of heat and pressure. Draw a second dark rectangle within the brackets showing how you think the force may alter the shape of the crust.

1. [ ]

5. 

2. 

6. 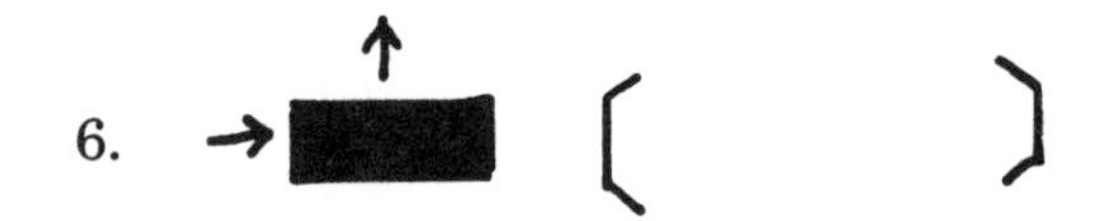

3. 

7. [ ]

4. [ ]

8. [ ]

When the Earth's surface is subjected to uplift, subsidence, or thrust, three changes may occur: folding, fracturing, or faulting. Folding is the upward bend in rock strata; fracturing refers to breaks in rocks due to folding and faulting; faulting happens when rock movement occurs along either side of a crack in the Earth's crust.

## PART 2

On the back of this sheet, write which of the changes—folding, fracturing, or faulting—you think might occur in the crust as indicated by the direction of the arrow in Part 1. You may decide that two or maybe all three changes may occur. Be sure to give a reason for your answer.

Name ______________________________ Date ____________________

# 9-10 Pancake Earth

**Introduction:** When the Earth's rocks undergo heat and pressure, several things happen: Rocks expand, contract, warp, buckle, and twist. Unquestionably, the Earth remains unsettled. Volcanic eruptions and earthquakes help our planet find relief from the forces of stress and strain within its crust.

**Objective:** To demonstrate the effect of heat on cooler substances and the effect of cooling on warmer substances.

**Materials:** Copies of Observation Charts 1 and 2, pancake flour or Bisquick™, Pyrex™ or metal tray, teaspoon, water, thermometer (F°), hot plate, magnifier, metric ruler, and beaker (100 ml).

**Procedure:** Do the following:

### *A. Part 1: The Effect of Heat on Cooler Substances*

- Mix 4 teaspoonsful of pancake flour with water in a 100-ml beaker. Add enough water to get a pasty substance. Stir until the flour mixes thoroughly with the water.
- Pour the mixture into a Pyrex™ dish or metal tray. Spread around in a circle. Place a thermometer (F°) in the mixture and record the temperature. Measure (mm) and record the diameter of the mixture. Place these recordings in the space provided on Observation Chart 1.
- Transfer the tray of doughy substance to a hot plate. Set the temperature dial at 350°F.
- Every two minutes for the next ten minutes, record on Observation Chart 1 what changes take place in the mixture. Check the appropriate places on the chart.

### *B. Part 2: The Effect of Cooling on Warmer Substances*

- After ten minutes, remove the mixture from the hot plate. Continue to observe dough temperature and texture changes every two minutes for twenty minutes. Make appropriate checks on Observation Chart 2. Allow the mixture to stand overnight. The following day, make final observations of structural and temperature changes.

### *C. Part 3: After 24 Hours*

- Once more, measure (mm) the length and width of the cracks previously measured in Part 2. Have they changed in size? If so, how? ______________________________
- Does this activity provide evidence showing that cooling of material in the Earth's crust causes contraction? ______________________________

  What does this activity show regarding the cooling of material? ______________________________

  ______________________________________________________________

- Has the mixture's diameter changed over the past 24 hours? If so, what does this indicate?

______________________________

______________________________

### *Questions*

Answer the following questions for Part 1 observation:

1. How does heat affect the mixture? ______________________________
2. What was the mixture's temperature after 10 minutes of heating? __________ °F. How much did it increase over the starting temperature? __________ °F.
3. When do bubbles appear in the mixture? ______________________________
4. When does the surface begin to dry? ______________________________
5. Describe how the following miniature structures form in the mixture:

   plateaus ______________________________

   plains ______________________________

   ocean basins ______________________________

   mountains ______________________________

   volcanoes ______________________________

   faults ______________________________

6. When does each of the structures listed in Question 5 appear (minutes)?

   plateaus __________ minutes

   plains __________ minutes

   ocean basins __________ minutes

   mountains __________ minutes

   volcanoes __________ minutes

   faults __________ minutes

Answer the following questions for Part 2 observation:

1. How does cooling affect the mixture? ______________________________

   ______________________________

2. When did cracking occur in the mixture? ______________________________

   Measure (mm) the length and width of the largest cracks.

   Length ______________ Width ______________

3. In the space below, sketch the approximate positions of the cracks in the mixture.

4. When did the mixture's surface appear dry? ______________________________

5. How much did the temperature decrease after 20 minutes of cooling? __________ °F

6. Make up your own theory on how Earth structures develop based on temperature variation, heat, pressure, and so on.

   ______________________________

   ______________________________

   ______________________________

   ______________________________

   ______________________________

## Observation Chart 1
## The Effect of Heat on Cooler Substances

Structural Changes in Mixture
(Check ✓ appropriate spaces below)

| Time (Min.) | Temp. (F) | Gas Bubbles Present | Surface Becoming Dry | Cracks Forming | Folds Developing | Holes Forming | Rapid Rising & Falling of Dough | Rapid Hardening of Surface | Very Little Change |
|---|---|---|---|---|---|---|---|---|---|
| | | | | | | | | | |
| | | | | | | | | | |
| | | | | | | | | | |
| | | | | | | | | | |
| | | | | | | | | | |
| | | | | | | | | | |
| | | | | | | | | | |
| | | | | | | | | | |
| | | | | | | | | | |
| | | | | | | | | | |
| | | | | | | | | | |

Diameter of mixture before heating ________________ mm

## Observation Chart 2
## The Effect of Cooling on Warmer Substances

Structural Changes in Mixture
(Check ✓ Appropriate Spaces Below)

| Time (Min.) | Temp. (F) | Gas Bubbles Present | Surface Becoming Dry | Cracks Forming | Folds Developing | Holes Forming | Rapid Rising & Falling of Dough | Rapid Hardening of Surface | Very Little Change |
|---|---|---|---|---|---|---|---|---|---|
| | | | | | | | | | |
| | | | | | | | | | |
| | | | | | | | | | |
| | | | | | | | | | |
| | | | | | | | | | |
| | | | | | | | | | |

Diameter of mixture after heating and cooling ________ mm

### Overnight

| Time (Hrs.) | Temp. (F) | Holes Grew Larger | Shrinking Occurred | Cracks Increased in Width and Length |
|---|---|---|---|---|
| 24 | | | | |

Diameter of mixture after 24 hours ________ mm

# Section 9: Forces That Shape the Earth's Surface
# TEACHER'S GUIDE AND ANSWER KEY

## 9-1 Restless Earth Term Puzzle

*Across:* 2. fault; 4. diastrophism; 7. elastic; 9. crust; 11. anticline; 14. earthquake; 15. compression; 16. thrust; 17. isostasy; 18. magma; 20. convection; *Down:* 1. uplift; 3. twist; 5. syncline; 6. fissure; 8. volcano; 10. tension; 12. fracture; 13. shearing; 18. mantle; 19. fold

## 9-2 Erosion at Work

### *Part 1*

1. talus; 2. silt; 3. delta; 4. dunes; 5. arid; 6. wind; 7. water

| w | a | t | e | r | a | f | e | d | l | o | e | s | d | i | c | e |
|---|---|---|---|---|---|---|---|---|---|---|---|---|---|---|---|---|
| r | s | a | e | m | b | w | r | u | e | r | o | d | e | h | p | a |
| a | a | l | t | e | r | i | o | n | o | i | l | s | l | u | m | p |
| i | n | u | e | s | a | n | d | e | b | a | s | h | t | k | a | c |
| n | d | s | g | a | l | d | e | s | i | l | t | e | a | r | i | d |

### *Part 2*

1. plucking; 2. exfoliation; 3. loess; 4. moraine; 5. abrade; 6. potholes; 7. alluvial fan; 8. rainstorm; 9. windbreak; 10. glaciers

## 9-3 Weathering Processes

The sketches and labels will vary. The ones shown here are samples.

1. Physical weathering. Rock broke without any change in the rock materials. Sample sketch:

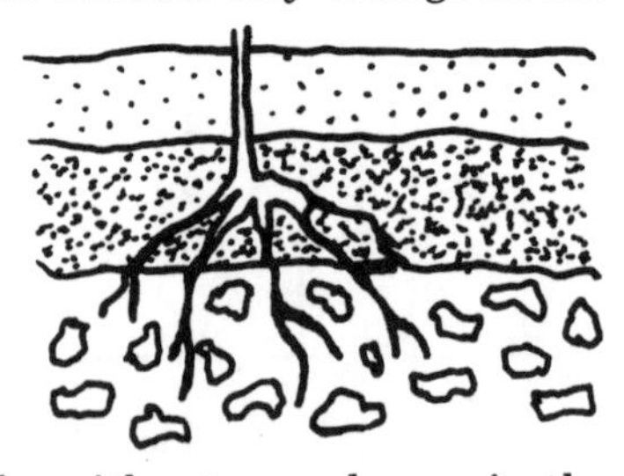

2. Physical weathering. Rock broke without any change in the rock materials. Sample sketch:

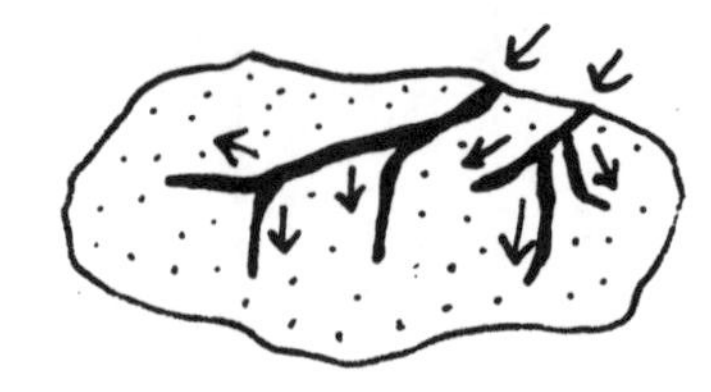

3. Chemical change (weathering). Some of the rock materials changed. Sample sketch:

4. Physical weathering. Rock broke without any change in the rock materials. Sample sketch:

## 9-4 Forces Around Us, Part 1

### *Part 1*

1. chemical weathering; 2. erosion; 3. joints; 4. V-shaped valleys; 5. earthquake; 6. meander; 7. delta; 8. batholith

### *Part 2*

*Possible answers*:

1. Large boulders may crumble to produce rock debris. Thus, a large buildup of materials develops in certain areas.
2. A gradual wearing away of parts of the Earth's crust
3. Splitting of rocks may add to the buildup of debris in certain areas.
4. Valleys cut their way into the Earth's crust.
5. Various parts of the Earth's crust buckle, twist, and stretch. Streams and rivers have had their courses changed, physical structures have suffered total destruction, and so on.
6. A river bed widens and forms a snake-like pattern.
7. A steady buildup of river deposits may form islands.
8. Huge boulders protrude out of the ground over a large area.

## 9-5 Forces Around Us, Part 2

*Possible answers*:

1. Weathering deposits may be used for landscaping, rock fill, or, in some cases, fertilizer or planting soil.
2. Erosion may destroy valuable farming land.
3. May weaken the ground over which buildings rest
4. Erosion may destroy valuable farming land or land needed for housing.
5. Destruction of property and buildings; injury or death to animals and people

6. Erosion may destroy valuable farming land or land needed for housing.
7. Provides new recreational areas; may also render some land useless for further development
8. May restrict building in rocky areas; may severely limit the growing of crops, etc.

## 9-6 Glaciers on the Move

### *Part 1*

1. valley glacier; 2. moraine; 3. crevasse; 4. cirque; 5. glacial; 6. kettle; 7. névé. The force that acts on the ice and causes it to flow is gravity.

### *Part 2*

drumlin—glacial hill; horn—sharp peak; cirque—steep basin; crevasse—glacial fissure; arête—sharp ridge; striations—parallel scratches; moraine—glacial till; firn—rough ice; valley—alpine glacier; kettle—circular hollow

### *Part 3*

1. By eroding the land, carving out valleys; 2. Provide beautiful landscapes, offer skiing in certain areas; 3. Erode landscape, providing year-round freezing temperatures; 4. You develop an understanding of nature's powerful forces at work, e.g., pressure, gravity, and agents of erosion.

## 9-7 Sleuthing the Newspaper

Chart information will vary. This is an excellent assignment for students to do over three or four days. It helps them become aware of the problems many people face all over the world. You might want to save several papers and have them on hand for those students who have trouble acquiring a newspaper.

## 9-8 Changing Earth Word Search

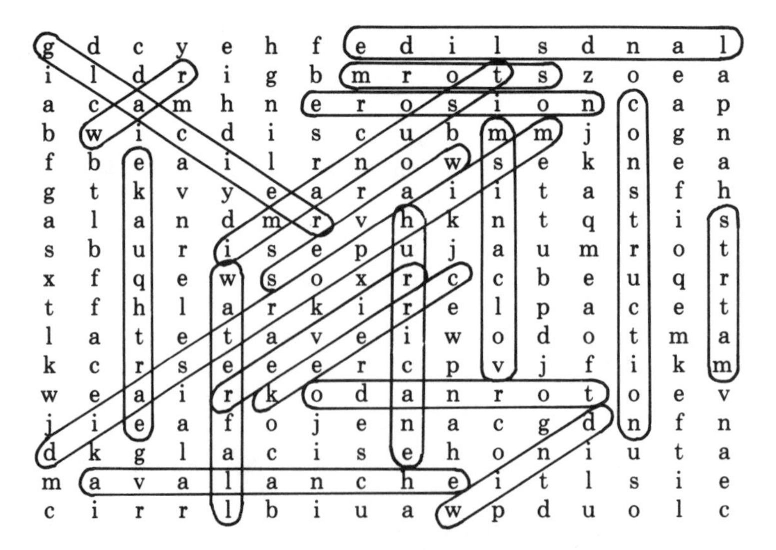


*Nature* (blue)—wind, glacier, earthquake, hurricane, tornado, tsunami, volcanism, storm, diastrophism, waves; *Partially created by humans* (green)—stream, river, creek, landslide, avalanche, erosion, waterfall; *Totally caused by humans* (red)—war, construction

## 9-9 Diastrophic Events

Sketches, answers, and reasons will vary.

## 9-10 Pancake Earth

Be sure to warn students that the Earth is not a hot plate, and that the crust is *not* like dough.

### *Part 3*

Possible answers: The cracks may have shrunk in length and width; some may not have changed since they were measured in Part 2.

Possible answers to Observation 1: 1. Heat causes the mixture to expand. Items 2–6: Answers will vary. Observation 2: 1. Cooling causes the mixture to shrink. Items 2–6: Answers will vary. Answers to Observation Charts 1 and 2 will vary.

## Section 9: Forces That Shape the Earth's Surface
## MINI-ACTIVITIES

Listed below are nine mini-activities, one to five minutes in length, for students to do at the beginning or end of the period.

1. **Riddle:** During a recent political campaign, Mayor Jones gathered his supporters together for a last rally before they went to the polls. He held the rally at his ranch in the foothills. As the crowd shouted and cheered, a large mass of loose bedrock came sliding down a nearby hill. Mayor Jones threw his hands in the air and hollered to the crowd, "If we could vote today, I'd . . ." What did he say?

   (**Answer:** "If we could vote today, I'd win by a landslide.")
2. Decode the following statement:

   (Raw + ks) + (w + [ear] ) + ah + way) + (bi or buy) +

   (grrr + [eye] + d + ing) + (act + sh + opposite of "off")

   (**Answer:** Rocks wear away by grinding action.)
3. Make up silly rhyming first and last names for boys and girls from the following glacier structures: moraine, firn, horn, cirque, till, esker, and kettle. Example: Lorraine Moraine.
4. Unscramble the earth-shaping terms below and fit them into the puzzle.

   **Across**

   2. olfd (fold)
   4. wgeniarteh (weathering)
   5. igrcale (glacier)

   **Down**

   1. seonori (erosion)
   3. atulf (fault)

5. **Riddle:** Why are some mountains like some athletes?

   (**Answer:** There are times when both fold under pressure.)
6. What do the following symbols and signs represent?

   $\equiv H_2O$

   (**Answer:** running water)

   | FORECAST:<br>Continued Showers |
   |---|

   (**Answer:** moraine)

ROCK

(**Answer:** erosion or weathering)

7. **Question:** How is runoff—water that flows over the Earth's surface—like a television set?
   (**Answer:** It has more than one channel.)
   **Question:** How is the Bank of America like a riverbank?
   (**Answer:** Both collect deposits.)
8. Shade in the squares that *do not* spell the answer to each description below.

   A downward fold in a rock
   (**Answer:** syncline)

| t | s | n | y | c | n | l | c | i | o | l | i | p | n | y | e |
|---|---|---|---|---|---|---|---|---|---|---|---|---|---|---|---|

   A large mass of moving ice and snow
   (**Answer:** glacier)

| r | l | g | c | l | i | a | p | c | a | i | r | e | n | r |
|---|---|---|---|---|---|---|---|---|---|---|---|---|---|---|

   The chemical breakdown of rock particles
   (**Answer:** decomposition)

| e | d | c | e | c | o | p | m | o | p | i | o | i | s | i | e | o | t | i | n | o | n |
|---|---|---|---|---|---|---|---|---|---|---|---|---|---|---|---|---|---|---|---|---|---|

9. List at least four earth-shaping forces that contain the letters "i" and "o."

   a. ____________________
   b. ____________________
   c. ____________________
   d. ____________________

   (**Answers:** a. erosion; b. volcanism; c. exfoliation; d. diastrophism; the answers can appear in any order.)

Section 10

# WEATHER and CLIMATE

## OUTLINE

10-1 Weather Word Search
10-2 Mixed-up Weather Term Puzzle
10-3 Weather Terms Coded Message
10-4 Tools of the Meteorologist
10-5 Fronts
10-6 Weather Trivia
10-7 Climate Vocabulary Fill-ins
10-8 Water Thermometer, Part 1
10-9 Water Thermometer, Part 2
10-10 Measuring Humidity with Hair
10-11 A Straw Barometer
10-12 Making a Wet- and Dry-Bulb Thermometer

## MATERIAL

The following laboratory materials are needed for Section 10:

- apron
- beakers (250 ml)
- beakers (1000 ml or larger)
- Bunsen burners
- cotton, cloth, or shoelace wicks
- Erlenmeyer flasks
- food coloring
- glass tubing (various lengths)
- glue
- goggles
- heat sources (alcohol lamps or candles)
- human hair
- ice
- index cards, 3″ by 5″
- masking tape
- metric rulers
- one-hole stoppers
- paper towels
- plastic beads
- quart jars or large-mouth containers
- ringstand and clamps
- soap
- stiff paper or cardboard
- straws
- thermometers
- thin rubber layers
- thread
- water
- wooden splints

Name ______________________ Date ______________________

# 10-1 Weather Word Search

Locate and circle the 20 terms commonly found in earth science textbooks related to weather. The terms are listed below the puzzle. They may be found backward, forward, vertically, horizontally, and diagonally.

```
a e t s i g o l o r o e t e m a p
w e s t e r l i e s e n p o i w s
r h u r r i c a n e o q v s l a y
e w e a t h e r x r a p i w k r c
t r w t y j g e f a o a t s s m h
e z c u e b t d o l d r u m s f r
m a f s r z e i a e p r e d a r o
o h u m i d i t y e r f n h m o m
m a e c u b c h n i g i m e r n e
e i v l t a i o c j w a o g i t t
n l c a o n z i d e i k e u a c e
a c y c l o n e b a r o m e t e r
o e i s a x w n o i t a i d a r e
```

| | | |
|---|---|---|
| air mass | hail | radiation |
| anemometer | humidity | stratus |
| barometer | hurricane | warm front |
| cirrus | meteorologist | weather |
| cyclone | occluded front | westerlies |
| dew | ozone | wind |
| doldrums | psychrometer | |

Name ______________________________ Date ______________________________

# 10-2 Mixed-up Weather Term Puzzle

## PART 1

Unscramble the terms in parentheses. One term matches the description. Write the term in the space provided. If no answer is given, write *not given. Hint:* Two answers are not given.

1. The atmospheric layer closest to the Earth: ______________________________
   (esrterhaptos, etrreohpops, not given)
2. The amount of water vapor in the air: ______________________________
   (yhtumidi, dsaettuar, not given)
3. The study of the atmosphere: ______________________________
   (ygegolo, ehyrdroeshp, not given)
4. The boundary between two air masses: ______________________________
   (ecnyolc, tfnro, not given)
5. The temperature at which air becomes saturated with water vapor: ______________
   (erveilat topni, edw tpino, not given)
6. An example of precipitation is ______________________________.
   (tlese, dnwi, not given)
7. Water changes from a liquid to a gas. ______________________________
   (nsaotitura, envoaitpoar, not given)
8. A wispy, curled type of cloud: ______________________________
   (sartuts, mclusuu, not given)
9. Intense storms formed over warm areas of the sea; they are given people's names.
   ______________________________ (otdoranse, sehnuracri, not given)
10. A scientist who studies the weather: ______________________________
    (tbsioiglo, tmestiegoolor, not given)

## PART 2

Match the description in the left-hand column with the term or example in the right-hand column. Write the letter of the term or example in the space next to the description.

| | | |
|---|---|---|
| ______ | 1. moisture | a. boundary |
| ______ | 2. cloud | b. westerlies |
| ______ | 3. wind belt | c. nitrogen |
| ______ | 4. ozone | d. stratosphere |
| ______ | 5. jet stream | e. humidity |
| ______ | 6. front | f. ultraviolet |
| ______ | 7. atmosphere | g. cirrus |
| ______ | 8. Coriolis effect | h. rotation |

Name ______________________________ Date ______________________

# 10-3 Weather Terms Coded Message

Each of the terms described below has the same code as the secret message. Fill in the letters for each term and reveal the coded message.

1. Piled-up, billowy clouds

   ___ ___ ___ ___ ___ ___ ___
   1 26 7 26 8 26 22

2. Makes up 78% of the atmosphere

   ___ ___ ___ ___ ___ ___ ___ ___
   11 15 18 6 3 12 10 11

3. All weather occurs in this atmospheric layer

   ___ ___ ___ ___ ___ ___ ___ ___ ___ ___ ___
   18 6 3 2 3 22 2 20 10 6 10

4. The atmospheric layer that protects the Earth from ultraviolet rays

   ___ ___ ___ ___ ___ ___ ___ ___ ___ ___
   7 10 22 3 22 2 20 10 6 10

5. Rising currents of warm air around the equator

   ___ ___ ___ ___ ___ ___ ___ ___
   13 3 8 13 6 26 7 22

6. The daily changes of pressure, humidity, temperature, and wind

   ___ ___ ___ ___ ___ ___ ___
   23 10 5 18 20 10 6

7. Condensed moisture suspended in air molecules—for example, stratus and cirrus

   ___ ___ ___ ___ ___ ___
   1 8 3 26 13 22

8. Small whirls of air that move up and down rapidly

   ___ ___ ___ ___ ___ ___ ___ ___ ___ ___
   18 26 6 4 26 8 10 11 1 10

9. Lines drawn on a map showing the same average yearly temperature

   ___ ___ ___ ___ ___ ___ ___ ___
   15 22 3 18 20 10 6 7

10. Air moving from the land toward the water (two words)

    ___ ___ ___ ___    ___ ___ ___ ___ ___ ___
    8 5 11 13    4 6 10 10 19 10

11. Air moving from the sea toward the land (two words)

    ___ ___ ___    ___ ___ ___ ___ ___ ___
    22 10 5    4 6 10 10 19 10

12. The ratio between the amount of water vapor actually in the air and the amount the air could hold at the same temperature (two words)

6 10 8 5 18 15 25 10    20 26 7 15 13 15 18 30

13. Lines drawn on a map having equal atmospheric pressure

15 22 3 4 5 6

14. A short-term prediction of weather patterns

9 3 6 10 1 5 22 18

15. An instrument that measures wind velocity

5 11 10 7 3 7 10 18 10 6

16. A measure of air pressure

7 15 8 8 15 4 5 6

17. Two air masses meet; the warm air replaces the cold air (two words)

23 5 6 7    9 6 3 11 18

18 Two air masses meet; the cold air replaces the warm air (two words)

1 3 8 13    9 6 3 11 13

*Secret Message:* 22 18 6 5 18 26 22    1 8 3 26 13 22 9 3 6 7

15 11 8 5 30 10 6 22.

Name ______________________ Date ______________________

# 10-4 Tools of the Meteorologist

## PART 1

Use the combination of words, phrases, and sketches below to identify the tools used by meteorologists to forecast weather conditions. Write your answers in the space provided.

1. centigrade + 212°F + mercury + alcohol + measurement of air temperature

   Answer: ______________________

2. contains no liquid + sealed metal container + air pressure + 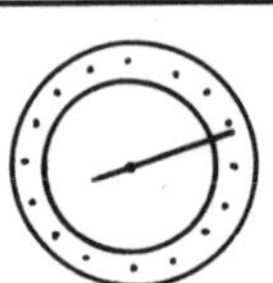

   Answer (two words): ______________ ______________

3. track storms + short-wave radio +  + storms appear as bright spots

   Answer: ______________________

4. cameras to photograph cloud patterns + orbit + provide information for weather maps

   Answer (two words): ______________ ______________

5. rain or snow + scale + funnel + weighing scale

   Answer (three words): __________ __________ __________

6.  + weather symbols + fronts + forecasting

   Answer (two words): ______________ ______________

7. 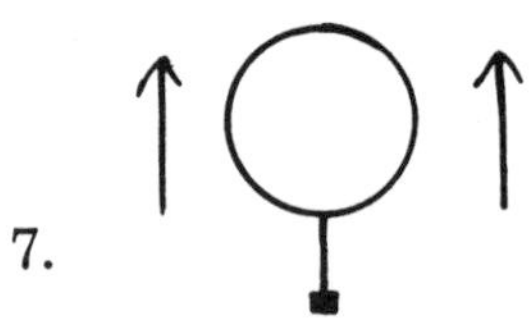 + "sound" + measures relative humidity + radio transmitter

   Answer: ______________________

8. pen arm + 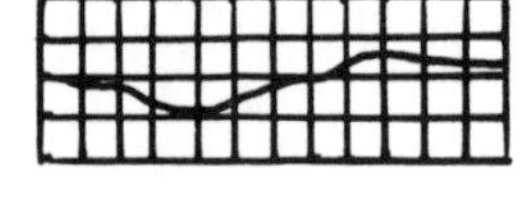 + air pressure + recording graph

   Answer: ______________________

## PART 2

On the back of this sheet, follow the same format as in Part 1 to describe any *two* of the following instruments used to forecast the weather: anemometer, mercurial barometer, psychrometer, or thermograph. Also, tell the function of each instrument.

Name ______________________ Date ______________

# 10-5 Fronts

Using an earth science textbook for reference, sketch and label a cold front, warm front, and stationary front in the spaces below. Color the cold areas blue and the warm areas red. Below each sketch briefly describe how each front forms and the type of weather associated with this condition.

### Cold Front

A cold front forms when ______________________

______________________

The type of weather associated with a cold front is ______________________

______________________

### Warm Front

A warm front forms when ______________________

______________________

The type of weather associated with a warm front is ______________________

______________________

### Stationary Front

A stationary front forms when ______________________

______________________

The type of weather associated with a stationary front is ______________________

______________________

Name ________________________________ Date ____________________

# 10-6 Weather Trivia

Here's a chance for you to find out how well you know weather trivia. Answer the following questions related to weather.

1. What is the freezing point of water on the Celsius temperature scale? ____________
2. A maximum thermometer stays at the highest temperature reached. What does a minimum thermometer do? ______________________________________________
______________________________________________
3. If an isotherm is a line drawn on a weather map to show places with the same temperature, what is an isobar? ______________________________________________
______________________________________________
4. What does a falling barometer usually mean? ______________________
______________________________________________
5. What two things does the following weather symbol show?
______________________________________________
6. Why do sailing vessels try to avoid the doldrums? ______________________
______________________________________________
7. What high-pressure areas that lie at about 30° to 35° latitude both north and south of the equator carry an animal's name? ______________________
8. Do you know at what point dew forms? ______________________
9. What does the following weather symbol indicate about sky conditions? 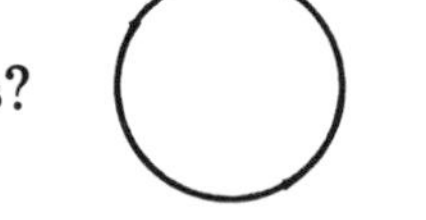
______________________________________________
10. What does the symbol cT mean? ______________________
11. What major feature separates a cold front from a stationary front? ____________
12. What is a willy-willy? ______________________
13. What do you call a tornado that passes over a body of water? ____________

14. What does "Mackerel scales and mares' tails make lofty ships carry low sails" mean?

______________________________________________

______________________________________________

15. Name the clouds responsible for causing a halo to appear around the sun or the moon.

______________________________________________

16. Why is a humid climate generally uncomfortable?

______________________________________________

______________________________________________

17. Why does the ozone layer contain high temperatures?

______________________________________________

______________________________________________

18. What do evaporation, condensation, and humidity have in common?

______________________________________________

Name ______________________________ Date ______________________

# 10-7 Climate Vocabulary Fill-ins

Complete these sentences by filling in each missing word from the word list.

1. A term meaning long-term weather: ______________________
2. ______________________ is a main feature of climate.
3. Another main feature of climate is ______________________.
4. The region around the equator is known as the ______________________ zone.
5. The polar regions are known as the ______________________ zones.
6. A region that has an average temperature of more than 68°F for the year is considered to have a ______________________ climate.
7. Tropical ____________ climate produces light rainfall and ____________ temperature.
8. A ______________________ climate produces very light rainfall, short, hot summers, and long, cold winters.
9. Regions that have an average temperature of less than 50°F during their warmest month are considered to have a ______________________ climate.
10. Regions bordering the tropical deserts; may extend into the middle latitudes: ____________
11. A ______________________ is a seasonal circulation of air masses from land to sea in winter and from sea to land in summer.
12. ______________________ regions show an annual rainfall of less than 25 cm.
13. Mountainous regions generally have ____________ climates than surrounding areas.
14. A ______________________ climate exists when an area is located within a large land mass; it has mostly hot summers and cold winters (two words).
15. A ______________________ climate exists when an area is near an ocean; it has mostly cool summers and mild winters.
16. Prevailing winds are classified as climatic controls. When they come from the ocean, they carry ______________________ air masses onto the west coast of the continent.
17. The Labrador ______________________ is a cold current that affects the temperature of surrounding areas.

18. Climates may be classified as ______________________ or ______________________.

19. Climates may also be described in terms of plant life, such as rain forests and ________ ______________________.

20. A ______________________ is the climate in a limited area—for instance, the climate under a rock or inside a hollow log.

## Word List

| | | | |
|---|---|---|---|
| arid | frigid | microclimate | subarctic |
| climate | hot | monsoon | temperature |
| continental | humid | polar | torrid |
| cooler | light | rainfall | tropical |
| current | marine | steppes | tundra |
| desert | maritime | | |



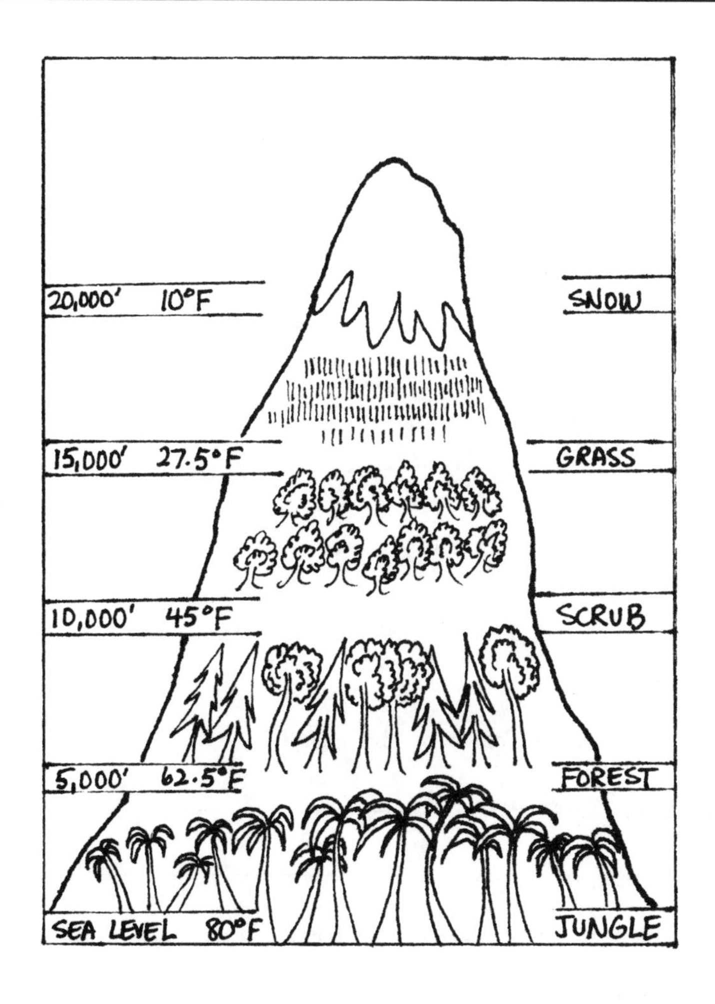

Name ______________________________ Date ______________________________

# 10-8 Water Thermometer, Part 1

**Introduction:** A thermometer measures the average speed of all moving molecules at a particular moment. This measurement indicates a change in temperature when a body adds or loses heat energy. Fast-moving molecules increase the temperature reading, and slow-moving molecules lower it. The expansion and contraction of molecules greatly influence weather conditions.

**Objective:** To build and test a colored-water thermometer.

**Materials:** Erlenmeyer flask, water, food coloring, Bunsen burner, glass tubing, one-hole stopper, ringstand and clamp, masking tape, goggles, and apron.

**Procedure:** Do the following:

1. Fill a flask with water. Add food coloring to the water.
2. Heat a length of glass tubing (15″ to 18″) until it can be bent into a U-shape.

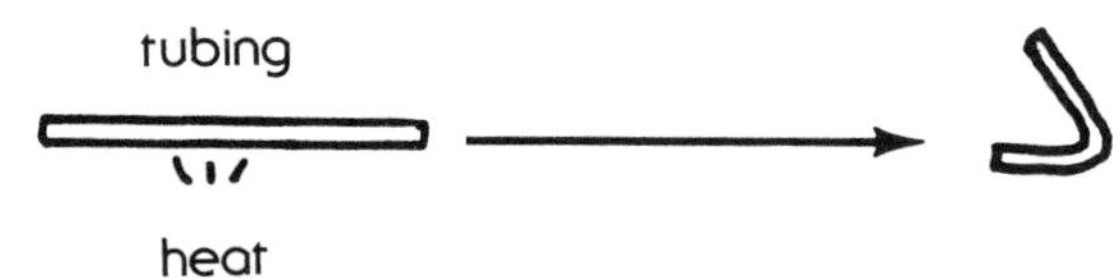


3. Carefully insert the bent tubing into a one-hole stopper. *Note:* Moisten the tubing and stopper with water or a soapy solution. This allows the tubing to slip through the stopper easily.
4. Press the stopper into the flask, invert, and support on a ringstand and clamp.

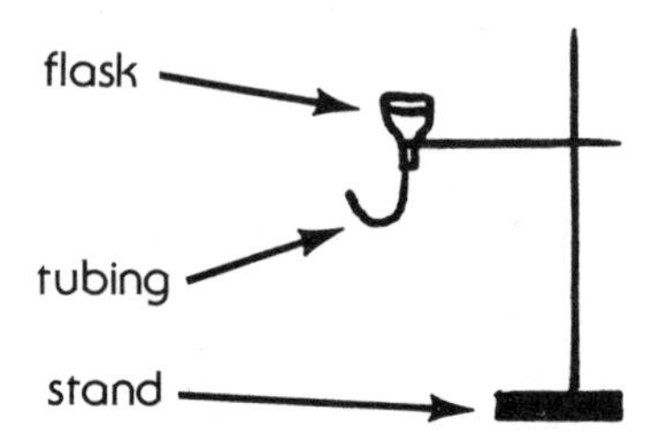


5. Observe the water level in the tube. Place a thin strip of masking tape on the tube over the water level.

## PART 1

- Now place your hand over the bottom of the flask.

  a. What happened to the water level? ______________________________

  ______________________________

  b. What caused the water level to do what it did? ______________________________

  ______________________________

## PART 2

- Remove your hand from the bottom of the flask.

  a. What happened to the water level? ______________________

  ______________________

  b. What caused the water level to do what it did? ______________________

  ______________________

  c. Does the activity measure the average speed of all moving molecules? Why or why not?

  ______________________

  ______________________

  ______________________

  d. Does the activity indicate a change in temperature when a body adds or loses heat energy? Why or why not? ______________________

  ______________________

  ______________________

  e. What does the activity show regarding the water level in the tube? ______________________

  ______________________

  ______________________

Name ______________________ Date ______________________

# 10-9 Water Thermometer, Part 2

**Introduction:** Refer to Activity 10-8: Water Thermometer, Part 1.

**Objective:** To build and test a colored-water thermometer.

**Materials:** Erlenmeyer flask, warm water, food coloring, Bunsen burner, long glass tubing (about 2 feet long), one-hole stopper, masking tape, large beaker (1,000 ml or larger), and ice.

**Procedure:** Do the following:

- Place a flask with colored water inside a large, empty beaker. Keep the neck of the flask above the water line.
- Insert a long glass tube into a one-hole stopper. Place the stopper on the flask.

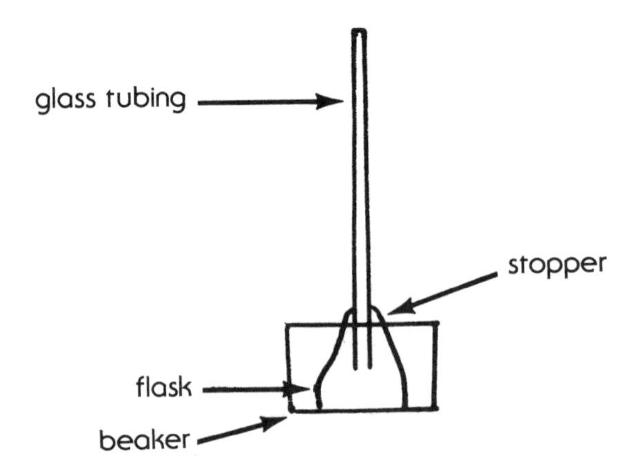


- Now test what effect, if any, the following conditions have on the water level in the tube.

1. Add ice water to the beaker. Effect? ______________________

______________________

2. Slowly pour hot water into the beaker. Effect? ______________________

______________________

3. Place warm, colored water into the flask. Submerge the flask into a beaker of ice water.

Effect? ______________________

______________________

4. Submerge a flask of colored ice water into a beaker of hot water. Effect? ______________________

______________________

Name ________________________________ Date ________________________

# 10-10 Measuring Humidity with Hair

**Introduction:** When people talk about humid weather, they're referring to the amount of moisture or vapor in the air. A person living in a humid region may experience discomfort because moisture in the air keeps perspiration from evaporating quickly. He or she might complain of feeling damp or sticky.

Moisture affects hair in an interesting way. You'll discover for yourself what hair does as you perform these experiments.

**Objective:** To test the ability of human hair to measure humidity.

**Materials:** Wooden splint, pencil, index card (3″ by 5″), thread, glue, plastic bead, metric ruler, human hair, soap, water, paper towel, ringstand, beaker (250 ml), and heat source.

**Procedure:** Do the following:

1. Fold the index card into thirds. Draw several straight lines, 2 millimeters apart, on the inside of the card (see the figure).
2. Set the card on a table or flat desk. Lay a wooden splint across the top of the card. Glue a plastic bead to a long hair. *Note:* Clean the hair with a soap solution. Wipe it dry.
3. Tie the opposite end of the hair to the center of the wooden splint. Set the bead at eye level with the center line on the card.
4. Check each day for humidity changes.

Answer these questions:

1. If humidity increases, what do you think will happen to the bead? Why do you think so?

______________________________________________________________

______________________________________________________________

2. If humidity decreases, what do you think will happen to the bead? Why do you think so?

______________________________________________________________

______________________________________________________________

3. Do you think measuring humidity with hair gives accurate results? Why or why not?

______________________________________________________________

______________________________________________________________

4. What actually happened to the hair as the humidity changed? ____________

______________________________________________________________

______________________________________________________________

Now boil water in a beaker. Set the hair instrument on a ringstand over the steam coming from the beaker. Does the steam change the position of the bead; that is, does the bead move up or down? ______________ Explain the results. ______________

_______________________________________________

_______________________________________________

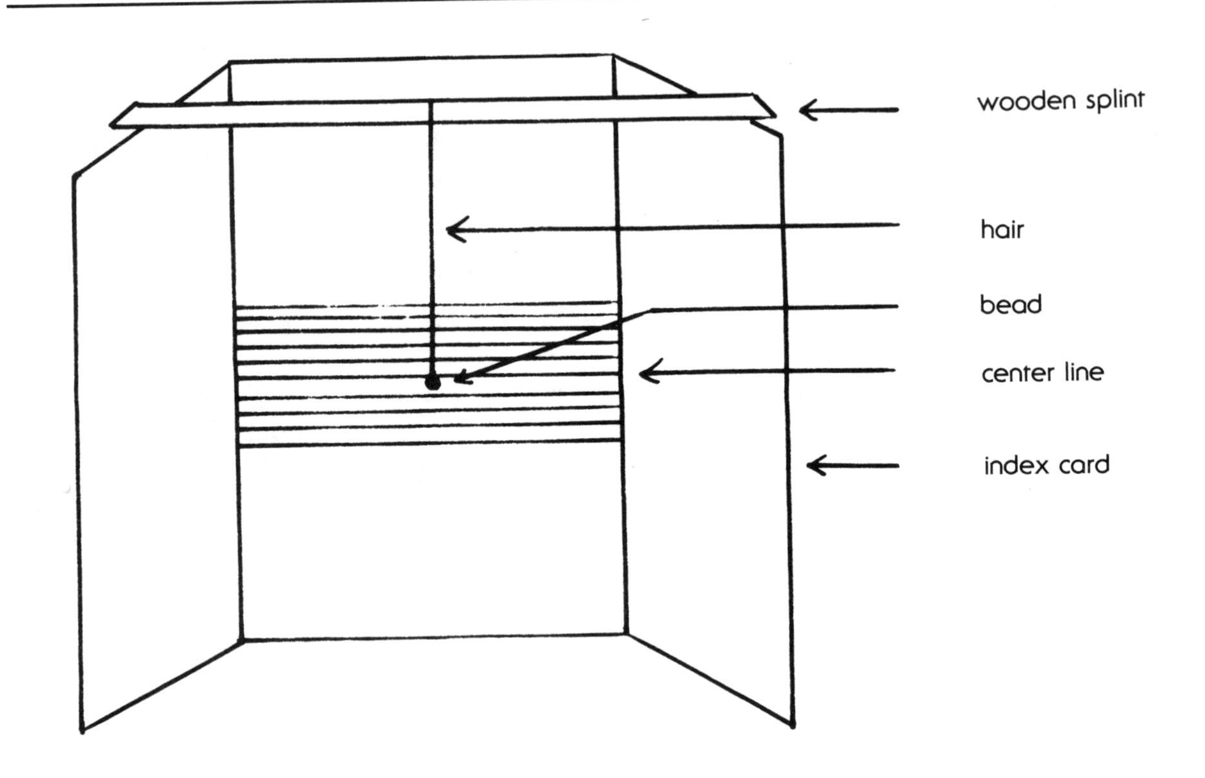

Name ______________________ Date ______________

# 10-11 A Straw Barometer

**Introduction:** Torrecelli, an Italian scientist, invented the barometer, an instrument for measuring atmospheric pressure. He found that mercury would rise approximately 30 inches in a tube free of air. He concluded that air pressure pushing down on the dish of mercury caused it to rise in the tube.

A rising barometer generally indicates good weather; a falling barometer suggests inclement weather.

**Objective:** To build and test a straw barometer.

**Materials:** Quart jar or large-mouth container, thin layer of rubber (large rubber balloon), glue, string, straw, and paper and pencil.

**Procedure:** Do the following:

1. Stretch and fasten with string a thin layer of rubber over the mouth of a jar.
2. Glue one end of a strand of straw to the center of the rubber covering.
3. Move the straw barometer to several areas around your school. Take daily readings, and complete the chart below. Wherever you set the barometer, make sure you mark a point directly behind the straw to indicate movement.

| Area | Date | Location | Present Weather Conditions | Straw Movement (Up, Down, No Movement) |
|---|---|---|---|---|
| 1 | | | | |
| 2 | | | | |
| 3 | | | | |
| 4 | | | | |
| 5 | | | | |

Answer these questions:

1. If the straw points upward, what can you say about the air pressure? ______________
2. If the straw points downward, what can you say about the air pressure? ______________
3. What does a barometer actually measure? ______________

______________
4. Are barometers intended to tell the nature of coming weather? Why or why not? ______

______________

______________
5. List two reasons a straw barometer is not a precision instrument.

a. ______________

b. ______________

Name ______________________________ Date ______________________

# 10-12 Making a Wet- and Dry-Bulb Thermometer

**Introduction:** Humidity, the amount of moisture in the air, can be measured by a wet- and dry-bulb thermometer known as a psychrometer. It operates on the principle of evaporation: The faster the evaporation, the cooler the temperature, and the lower the reading on the thermometer. If evaporation is slow, the cooling effect is less.

Two identical thermometers measure relative humidity. One thermometer has a cloth, cotton, or shoelace wick tied to its base. The tied material extends into a small dish of water. This is the wet-bulb thermometer. The other thermometer (dry bulb) is a regular thermometer for measuring air temperature.

Water rises from the dish and moistens the wick. Cooling of the wet-bulb thermometer by fanning it with a piece of paper or card causes moisture to evaporate from the wick. The relationship between wet and dry temperatures indicates relative humidity.

Relative humidity is an expression, written as a percent, of the amount of water actually present in the air compared to the maximum amount of water vapor the air can hold at that temperature.

**Objective:** To build and test a wet- and dry-bulb thermometer.

**Materials:** Two thermometers, cotton, cloth, or shoelace wicks, string, ringstands, and stiff paper or cardboard.

**Procedure:** Do the following:

1. Attach a single-layer wick around the bulb of one thermometer. Dip the wick in a small dish of water.

2. Hang both thermometers from a ringstand. Keep them 2 or 3 inches apart.

3. Fan both thermometers with a stiff piece of paper or cardboard. Stop fanning the thermometers when the wet-bulb thermometer reading remains steady.

4. Record the readings from both thermometers.

5. Find the relative humidity. Subtract the temperature readings on the wet- and dry-bulb thermometers.

6. Check your readings against a relative humidity chart supplied by your teacher. Find the degree of difference between the two thermometers. Then locate the dry-bulb thermometer reading on the left side of the chart and the number directly in line with both of these readings. This will give the percent of relative humidity. For example, if the temperature difference is 14° and the dry bulb reads 72°, the relative humidity is 42%.

7. Now test the relative humidity for each of the following areas listed on the chart. Write your readings on the chart.

| Area | Area Location | Dry-Bulb Reading | Wet-Bulb Reading | Difference in Readings | Relative Humidity |
|---|---|---|---|---|---|
| 1 | classroom floor | | | | |
| 2 | laboratory table | | | | |
| 3 | near the ceiling | | | | |
| 4 | inside hall or passageway | | | | |
| 5 | outside the classroom | | | | |

Answer these questions:

1. Can the model of the wet- and dry-bulb thermometer be an accurate instrument? Why or why not? ____________________

____________________

2. What does a high relative humidity indicate? ____________________

____________________

3. What does a low relative humidity indicate? ____________________

____________________

# Section 10: Weather and Climate
# TEACHER'S GUIDE AND ANSWER KEY

## 10-1 Weather Word Search

| a | e | t | s | i | g | o | l | o | r | o | e | t | e | m | a | p |
|---|---|---|---|---|---|---|---|---|---|---|---|---|---|---|---|---|
| w | e | s | t | e | r | l | i | e | s | e | n | p | o | i | w | s |
| r | h | u | r | r | i | c | a | n | e | o | q | v | s | l | a | y |
| e | w | e | a | t | h | e | r | x | r | a | p | i | w | k | r | c |
| t | r | w | t | y | j | g | e | f | a | o | a | t | s | s | m | h |
| e | z | c | u | e | b | t | d | o | l | d | r | u | m | s | f | r |
| m | a | f | s | r | z | e | i | a | e | p | r | e | d | a | r | o |
| o | h | u | m | i | d | i | t | y | e | r | f | n | h | m | o | m |
| m | a | e | c | u | b | c | h | n | i | g | i | m | e | r | n | e |
| e | i | v | l | t | a | i | o | c | j | w | a | o | g | i | t | t |
| n | l | c | a | o | n | z | i | d | e | i | k | e | u | a | c | e |
| a | c | y | c | l | o | n | e | b | a | r | o | m | e | t | e | r |
| o | e | i | s | a | x | w | n | o | i | t | a | i | d | a | r | e |

## 10-2 Mixed-up Weather Term Puzzle

### *Part 1*

1. troposphere; 2. humidity; 3. not given; 4. front; 5. dew point; 6. sleet; 7. evaporation; 8. not given; 9. hurricanes; 10. meteorologist

### *Part 2*

1. e; 2. g; 3. b; 4. f; 5. d; 6. a; 7. c; 8. h

## 10-3 Weather Terms Coded Message

1. cumulus; 2. nitrogen; 3. troposphere; 4. mesosphere; 5. doldrums; 6. weather; 7. clouds; 8. turbulence; 9. isotherm; 10. land breeze; 11. sea breeze; 12. relative humidity; 13. isobar; 14. forecast; 15. anemometer; 16. millibar; 17. warm front; 18. cold front

*Secret Message:* Stratus clouds form in layers.

## 10-4 Tools of the Meteorologist

Have on hand several books that show pictures of meteorology instruments.

### *Part 1*

1. thermometer; 2. aneroid barometer; 3. radar; 4. weather satellite; 5. weighing rain gauge; 6. weather map ; 7. radiosonde; 8. barograph

### *Part 2*

Student creations will vary. The listed instruments have the following functions: anemometer—measures wind velocity; mercurial barometer—measures air pressure; psychrometer—measures relative humidity; thermograph—records temperature with time.

## 10-5 Fronts

Sketches will vary. A *cold front* forms when a colder air mass thrusts under a warmer air mass. The type of weather associated with a cold front is violent storms.

A *warm front* forms when a warmer air mass overrides a colder air mass. The type of weather associated with a warm front is generally rain and showers.

A *stationary front* forms when both cold and warm fronts come to a halt for several days. The type of weather associated with a stationary front is generally rain for several days.

## 10-6 Weather Trivia

1. Zero degrees Celsius.
2. Stays at the lower temperature it has reached
3. A line drawn on a weather map connecting places of the same atmospheric pressure
4. Warmer weather, humid air
5. Wind direction and speed
6. Doldrums often mean no wind; no wind, no sailing
7. Horse latitudes
8. Dew point
9. Clear; no clouds present
10. Continental tropical—a warm, humid air mass that comes from tropical seas
11. A cold front moves; a stationary front stays in one place.
12. The name Australians give to hurricanes
13. Waterspout
14. The cloud formations suggest there might be a cyclone brewing with strong winds and rain or snow approaching.
15. Cirrus clouds
16. Moisture-laden air keeps body heat from escaping and slows down the evaporation process.
17. Ozone absorbs ultraviolet rays from the sun.
18. All deal with water.

## 10-7 Climate Vocabulary Fill-ins

1. climate; 2. rainfall; 3. temperature; 4. torrid; 5. frigid; 6. tropical; 7. light, hot; 8. subarctic; 9. polar; 10. steppes; 11. monsoon; 12. desert; 13. cooler; 14. continental marine; 15. marine; 16. maritime; 17. current; 18. arid, humid; 19. tundra; 20. microclimate

## 10-8 Water Thermometer, Part 1

It might be a good idea to let students practice bending glass before they do the activity. Give them pieces of glass tubing, 6 to 8 inches long, with rubber stoppers to place at both ends. Have students hold the glass tubing at both ends by the stoppers and place the center of the glass tubing over the burner. Rotate the glass piece over the burner until it begins to bend. Then carefully guide the piece into an L-shaped or right-angle pattern. Remove it from the burner and set it aside to cool.

*Part 1*

a. It rises.
b. Heat from the hand caused the bottom of the flask to warm. The expanding warm air caused the water level to rise.

*Part 2*

a. It dropped.
b. The air in the flask cooled and caused the water level to recede.
c. No. It only shows that warm air causes expansion and cool air causes contraction of water level.
d. No. There are no markings or calibrations to indicate temperature.
e. It may rise, lower, or stay the same.

## 10-9 Water Thermometer, Part 2

1. Water level lowers; 2.Water level rises; 3. Water level drops; 4. Water level drops, then rises.

## 10-10 Measuring Humidity with Hair

1. The bead should drop below the center line. Moisture causes hair to stretch.
2. The bead should rise above the center line. The hair will shrink.
3. No. It only serves as an indicator to show the presence of humidity in the air.
4. Answers will vary.
   (page 212) Steam will cause the hair to stretch, thus the bead will drop down.

## 10-11 A Straw Barometer

Chart answers will vary.

1. High pressure pressing down on the rubber causes the straw to rise.
2. Low pressure might cause the straw to point downward; there may be no noticeable movement.
3. Changes in pressure force
4. No. Changes in air pressure do not guarantee a certain type of weather.
5. a. A straw barometer uses crude materials. b. There are no accurate means to measure the changes in air pressure.

## 10-12 Making a Wet- and Dry-Bulb Thermometer

Be sure to have on hand a supply of relative humidity charts. Remind students to handle the thermometers with care as they move about the room and outside the corridor.

1. Not really. This crude model is subject to error.
2. Much moisture in the air
3. Small amount of moisture in the air

# Section 10: Weather and Climate
## MINI-ACTIVITIES

Listed below are 11 mini-activities, one to five minutes in length, for students to do at the beginning or end of the period.

1. **Riddle:** What does the circled letter in the word HURR(I)CANE represent?
   (**Answer:** the eye of the hurricane)
2. Complete the puzzle with the names of types of clouds.

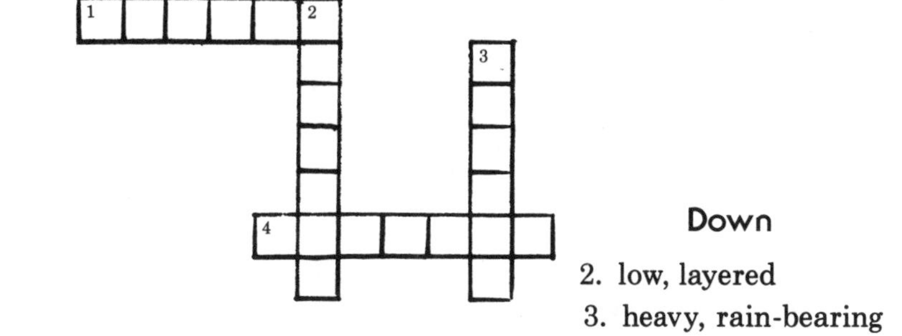

**Across**

1. thin, featherlike
4. thick, billowy

**Down**

2. low, layered
3. heavy, rain-bearing

   (**Answers:** Across—1. cirrus; 4. cumulus; Down—2. stratus; 3. nimbus)
3. Match Column A terms with the one-word descriptions in Column B. Draw a straight line connecting the term with the description.

| Column A | Column B |
|---|---|
| jet stream | condensation |
| fog | units |
| air mass | maritime |
| millibars | wind |

   (**Answers:** jet stream—wind; fog—condensation; air mass—maritime; millibars—units)
4. Name three cloud formations that have the letter combination "os" in their names.
   (**Answers:** cirrostratus, altostratus, and nimbostratus)
5. **Riddle:** What breakfast cereal do meteorologists prefer to eat?
   (**Answer:** Snow Flakes)
6. List as many people's names as you can think of that have something to do with weather.
   (**Possible answers:** Gale, Storm, Crystal, Willy, and so on)
7. Think of at least three first- and last-name combinations that a meteorologist might have. The sillier the better! Example: C. Breeze.
   (**Possible answers:** Wendy Day, Millie Bar, Anna Roid, Anna Momader, Faron Hite, Cindy Graid, Gale Winns, Harry Cane, and so forth)
8. Write the words related to climate that rhyme with the words given.

serene ____________ Havana ____________ molar ____________

horrid ____________ aptitude ____________ pontoon ____________

   (**Answers:** serene—marine; Havana—savanna; molar—polar; horrid—torrid; aptitude—latitude; pontoon—monsoon)

9. Match the weather symbols with the weather conditions by drawing a straight line between Column A and Column B.

| Column A | Column B |
|---|---|
| 1. | a. isobar |
| 2. | b. snow |
| 3. | c. wind from northwest at 15 knots, clear skies |
| 4. 1008 1008 | d. cold front, moving in a northwest direction |

(**Answers:** 1c, 2d, 3b, 4a)

10. Write at least two words after each letter in the word WEATHER that relate to weather or climate.

W ______________________, ______________________

E ______________________, ______________________

A ______________________, ______________________

T ______________________, ______________________

H ______________________, ______________________

E ______________________, ______________________

R ______________________, ______________________

(**Possible answers:** W—westerlies, wind; E—easterlies, energy; A—altitude, altostratus; T—thermometer, temperature; H—humidity, hurricane; E—electricity, equator; R—relative humidity, rain)

11. **Riddle:** What kind of weather is it when it's raining ducks and chickens?
(**Answer:** foul weather)

Section 11

# ASTRONOMY

## OUTLINE

11-1 Astronomy Word Search
11-2 Moon Term Game
11-3 Nebular Theory
11-4 Planet Match-up
11-5 What Goes with What?
11-6 How Do Planets Compare in Size (Diameter) with the Sun?
11-7 How About the Moon?
11-8 Constellations in a Minute!
11-9 Investigating a Mystery Planet
11-10 Meteorite Impact

## MATERIAL

The following laboratory materials are needed for Section 11:

- aluminum foil
- beakers (50–100 ml)
- burners
- cardboard
- clay
- cobalt chloride paper
- compound microscopes
- construction paper (black or blue)
- copy machine paper (teacher-prepared)
- drawing paper
- forceps
- glass tubing (different sizes)
- glue
- iron filings
- litmus paper
- magnets
- magnifying glasses
- metric rulers
- microscope slides
- mortars and pestles
- newspaper
- one-holed rubber stoppers
- paint
- paperclips
- paper matches
- paper towels
- pencils or pens
- plain white paper (8½″ by 11″)
- safety goggles
- scissors
- scoops
- small jars
- small stiff paintbrushes
- steel bearings (small, medium, large)
- test tubes
- thermometers
- tongs
- toothpicks, rounded

Name ______________________ Date ______________

# 11-1 Astronomy Word Search

Locate and circle the 30 terms in the puzzle related in some way to astronomy. The terms are listed below the puzzle. They may be found backward, forward, vertically, horizontally, and diagonally.

| | | | | | | | | | | | | | |
|---|---|---|---|---|---|---|---|---|---|---|---|---|---|
| t | e | n | a | l | p | y | r | u | c | r | e | m | o |
| a | s | a | e | i | e | n | o | z | o | s | t | a | r |
| l | b | y | f | c | r | a | n | u | l | g | x | r | b |
| c | a | e | a | l | o | b | e | p | a | i | e | a | i |
| t | i | p | x | r | g | m | a | l | s | b | s | l | t |
| s | s | g | n | c | e | n | e | u | e | b | a | o | a |
| u | j | u | p | i | t | e | r | t | m | o | h | s | e |
| n | l | a | n | i | e | b | r | o | e | u | p | m | g |
| e | a | r | t | h | s | a | o | s | s | s | e | a | a |
| v | e | a | c | o | s | n | p | a | b | t | l | r | n |
| a | n | e | m | a | t | i | t | s | e | a | e | i | o |
| i | t | i | u | a | l | u | a | o | x | c | i | a | r |
| g | e | q | w | c | r | g | r | y | e | p | a | f | o |
| d | w | a | e | n | i | s | e | b | c | s | e | r | c |

| | | |
|---|---|---|
| axis | lunar | Pluto |
| comet | maria | quasar |
| corona | Mars | rays |
| Deimos | Mercury | Saturn |
| Earth | meteor | solar |
| eclipse | moon | space |
| galaxy | orbit | star |
| gas | ozone | sun |
| gibbous | phase | Titan |
| Jupiter | planet | Venus |

Name ______________________________ Date ____________________

# 11-2 Moon Term Game

Write the word that best fits the description on the left. When you are finished, the boxed letters will answer the question: What did some early astronomers believe about the moon?

1. The first living creatures known to have visited the moon — _ _ _ _ [ ] _ _ _ _ _
2. Known to cover the moon's surface — _ _ _ [ ] _ _ _ _
3. The name of a large crater on the moon — _ _ [ ] _ _
4. A depression or hole left by a meteorite — _ _ _ _ [ ] _
5. When the moon is closest to the Earth — [ ] _ _ _ _ _ _
6. When the moon is farthest from the Earth — [ ] _ _ _ _ _
7. Moon's path in space — _ [ ] _ _ _
8. Comes just after the new moon — _ _ _ _ _ _ _ [ ]
9. Comes just after a full moon — _ _ _ _ [ ] _ _
10. The moon appears as a round disc in this phase. (two words) — [ ] _ _ _  _ _ _ _
11. The Earth or moon passing through the shadow of the other — _ _ _ _ _ _ [ ]
12. Dark areas on the moon's surface — _ [ ] _ _ _
13. A lunar body force; only one-sixth that of the Earth — _ [ ] _ _ _ _ _
14. The phase of the moon preceding gibbous (two words) — _ _ _ _ [ ]  _ _ _ _ _ _ _ _
15. One of the forms presented periodically by the moon — _ [ ] _ _ _

What did some early astronomers believe about the moon? It was _ _ _ _

_ _ _ _  _ _  _ _ _ _ _ _.

Name ______________________________ Date ______________________

# 11-3 Nebular Theory

A nebula is a cloud of gasses or dust in space. Early scientists—Immanuel Kant of Germany and Simon de Laplace of France—thought the sun and planets originated from a nebula. They believed that a huge twirling mass of hot clouds cooled, shrank, and began to spin faster. As pieces of matter broke away, planets and moons formed from the escaping rings of matter. The large central mass eventually became the sun.

## PART 1

Unscramble the 12 words below used to describe the Nebular Theory. Write the unscrambled term in the space provided.

1. latepn ______________________
2. ubaeln ______________________
3. tretma ______________________
4. tnisetics ______________________
5. goeritani ______________________
6. aclntre ______________________
7. lscudo ______________________
8. asgse ______________________
9. escpa ______________________
10. tlmaarei ______________________
11. cesepa ______________________
12. ksnhar ______________________



## PART 2

Now match the words with their descriptions below. Write the correct word in the space provided.

1. To come into existence: ______________________
2. The stuff of which anything is composed: ______________________
3. Anything that takes up space and has weight: ______________________
4. A non-self-luminous body of the solar system revolving around the sun: ______________________
5. A mass of visible vapors or particles floating in space: ______________________
6. To avoid or get away from: ______________________
7. Something that has decreased in size or shape: ______________________
8. An object located in the center: ______________________
9. A person who studies or investigates events or objects related to science: ______________________
10. A boundless, empty void in all directions: ______________________

11. A gaseous body with a cloud-like appearance: ______________________

12. Material that does not have a definite volume or shape (plural): ______________

## PART 3

In the space below, reconstruct the Nebular Theory using the five terms below. Show, with arrows, how Immanuel Kant and Simon de Laplace thought the solar system originated.

sun planets hot clouds nebula moons

Name ______________________ Date ______________________

# 11-4 Planet Match-up

## PART 1

Match the planet features in the left-hand column with the planet in the right-hand column. Write the number of each feature in the space provided. The same number may be used more than once.

| Planet Feature | Planet |
|---|---|
| 1. Surrounded by rings | Mercury ________ |
| 2. Has at least 16 moons | Venus ________ |
| 3. Has a banded, layered atmosphere | Earth ________ |
| 4. About 17 times more massive than Earth | Mars ________ |
| 5. Has one moon | Jupiter ________ |
| 6. Brightest of all planets | Saturn ________ |
| 7. 95% carbon dioxide atmosphere | Uranus ________ |
| 8. Found to have two polar caps | Neptune ________ |
| 9. Has a mass 1/400th that of the Earth | Pluto ________ |
| 10. Diameter: 12,756 km; temperature: 15°C | |
| 11. Temperature: −50°C; two moons | |
| 12. Year equals 88 earth days; no moons | |
| 13. Day: 17–24 hours; blue-green colored atmosphere | |
| 14. Red spot thought to be a giant atmospheric disturbance | |
| 15. Temperature: −230°C; year equals 248 years | |

## PART 2

Read the following fictional story. Use the hints in the story to identify the planet(s) that most likely fit the description.

*In 1987, the Satellite Wanderer was sent out into space to report data from planets in the solar system. After several years, something went wrong with the transmitter; scientists could pick up only faded bits of information. They managed, however, to match the information with the planet under investigation. See if you can discover what planet goes with the following messages.*

1. . . . surface . . . craters everywhere . . . craters well preserved . . . no atmosphere . . . temperature . . . daytime . . . may exceed 400° C

   Answer: ______________________

2. . . . hydrogen, helium, methane . . . nine rings circle . . . good closeup of moons Oberon and Ariel

   Answer: ______________________

3. . . . viewing bulge at equator . . . bright-colored bands of gases . . . Moon Io . . . evidence of active volcanoes . . . strong magnetic force

   Answer: ______________________________

4. Northern ice cap . . . made up of . . . frozen water and carbon dio . . . spot ice caps . . . old stream beds . . . salmon-colored skies

   Answer: ______________________________

5. . . . rock/ice rings . . . Titan . . . moon . . . larger than planet Mercury . . . may have solid core

   Answer: ______________________________

Name ______________________ Date ______________________

# 11-5 What Goes with What?

## PART 1

Match the two terms in the left-hand column with the description in the right-hand column. Draw a straight line connecting the term and the description.

| Terms | Description |
|---|---|
| 1. sky, heavens | a. lunar |
| 2. Orion, Taurus | b. inner planets |
| 3. explosion, gas | c. asteroid |
| 4. Saturn, Pluto | d. telescope |
| 5. rotation, revolution | e. constellations |
| 6. waves, energy | f. measurements |
| 7. Venus, Mars | g. ecliptic |
| 8. mirrors, light | h. celestial |
| 9. mare, crater | i. moon's phases |
| 10. stars, Milky Way | j. earth movements |
| 11. solid, small | k. outer planets |
| 12. parsecs, light-year | l. prominence |
| 13. waxing, waning | m. galaxy |
| 14. sun, path | n. spectrum |

## PART 2

One term does not belong in each of the following groups. Circle the unrelated term and tell why you feel it doesn't fit with the others. Write your answer in the space provided.

1. corona, photosphere, chromosphere, meteoroid

______________________

2. ellipse, mare, orbit, path

______________________

3. eclipse, umbra, albedo, moon

______________________

4. Neptune, Pluto, Mercury, Uranus

______________________

5. Steady State, "Big Bang," Static Universe, Cygnus

______________________________________________

6. triangular, spiral, elliptical, irregular

______________________________________________

7. maria, cepheid, T Tauri, cluster variable

______________________________________________

8. Syrtis Major, Red Spot, red color, "canals"

______________________________________________



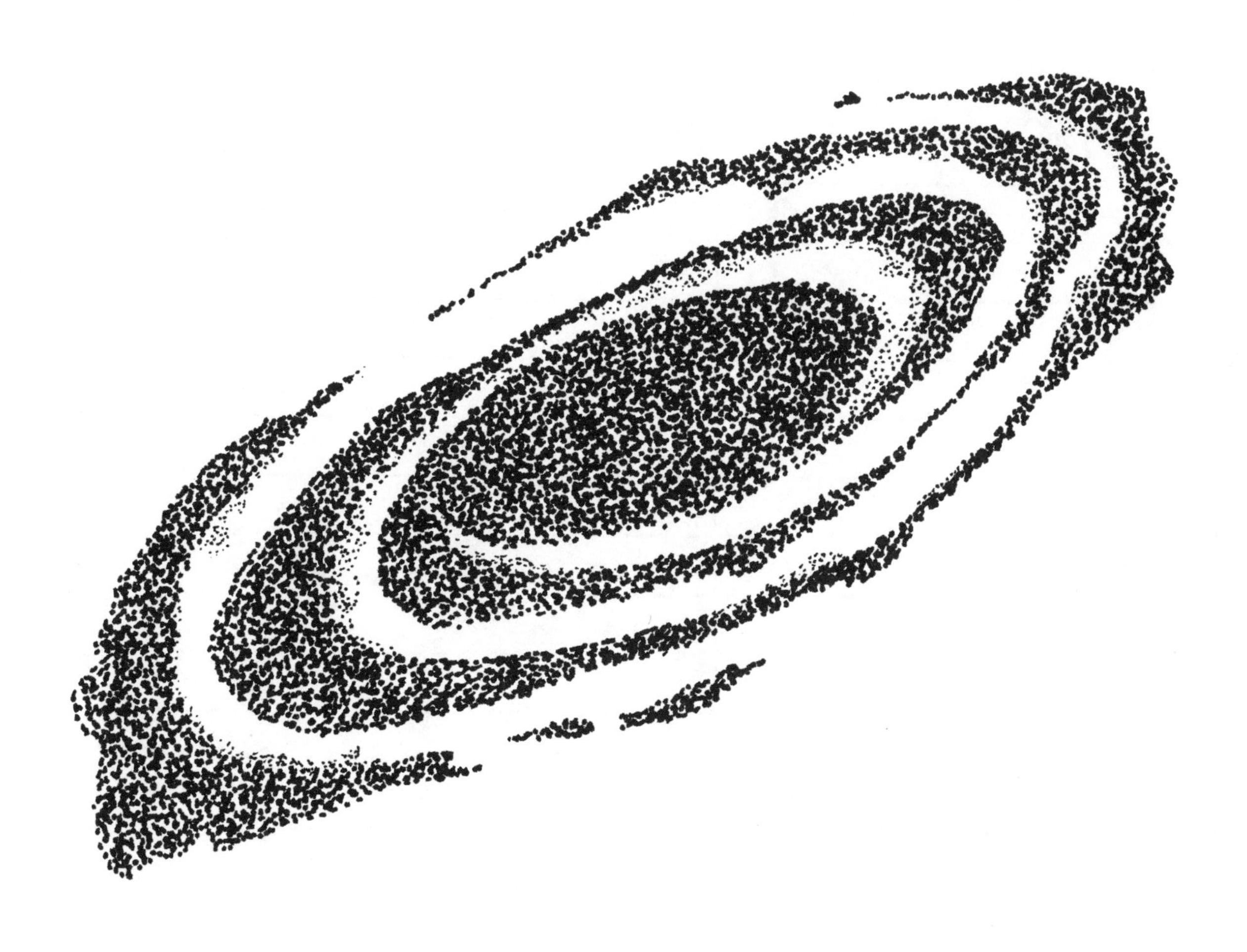

Name ______________________ Date ______________

# 11-6 How Do Planets Compare in Size (Diameter) with the Sun?

**Introduction:** The sun, a major source of heat and light, is a luminous body around which the Earth and other planets of the solar system revolve. A planet, then, is a heavenly body that revolves around the sun. Let's find out how planets compare in size (diameter) with the sun.

**Objective:** To compare each planet's diameter with the diameter of the sun.

**Materials:** Plain white paper (8½″ by 11″), construction paper, glue, scissors, and a metric ruler.

**Procedure:** Do the following:

## PART 1

- The chart shows each planet and its diameter in miles. Use the scale of 1 millimeter equals 10,000 miles to calculate the millimeter (mm) distance of each planet's diameter. Record your answers on the chart. *Hint:* Divide 10,000 into diameter in miles.

| Item | Diameter (Miles) | Millimeter (mm) Distance |
|---|---|---|
| Sun | 866,000 | |
| Mercury | 3,100 | |
| Venus | 7,700 | |
| Earth | 8,000 | |
| Mars | 4,200 | |
| Jupiter | 89,000 | |
| Saturn | 71,500 | |
| Uranus | 32,000 | |
| Neptune | 31,000 | |
| Pluto | 3,600 | |

- Measure and cut out each millimeter diameter from construction paper. Begin with the sun. *Note:* If some of the diameters are too small to cut out, make a dot with a pen or pencil to match the planet's diameter.
- Glue the sun to the center of a piece of plain white paper.
- Glue each planet around the sun according to its distance from the sun—for example, Mercury, closest to the sun; Venus, next closest to the sun.

## PART 2

When you finish making the model, examine your final product and answer these questions:

1. If all of the planets were placed side by side inside the sun, how much of the sun's diameter (mm) would be covered?

 ______________________________

2. How many Jupiters (mm) placed side by side would equal the sun's diameter (mm)?

 ______________________________

3. If Planet X had a diameter (mm) that equaled the total diameters (mm) of Saturn and Jupiter combined, how much larger would the sun's diameter be?

 ______________________________

4. If your space vehicle traveled 400 miles per hour, how many hours would it take to travel across the sun? How many days?

 Hours: ____________ Days: ____________

5. Write six true statements based on the results of this activity. For example: The sun's diameter is 9½ times larger than the diameter of Jupiter.

 a. ______________________________

 b. ______________________________

 c. ______________________________

 d. ______________________________

 e. ______________________________

 f. ______________________________

Name ______________________________ Date ____________________

# 11-7 How About the Moon?

Where did the moon come from? Some scientists believe that the moon may have had three possible origins:

1. At one time, the moon was part of the Earth. The rotation of the Earth coupled with the sun's gravity caused a bulge to form. The bulge eventually separated from Earth, flew into space, and became the moon. This is known as the daughter theory.
2. A twin theory suggests that the moon and Earth, two separate planets, formed at the same time. Scientists call this the sister planet theory.
3. The capture theory claims that the moon was formed in another part of the solar system. At some time, the Earth's gravity "captured" the moon.

## PART 1

Reconstruct each of the theories by sketching them in the space provided, using only the symbols found in the lower portion of the box. Do not label any part of the sketch. You can use a symbol more than once; some symbols may not be used at all.

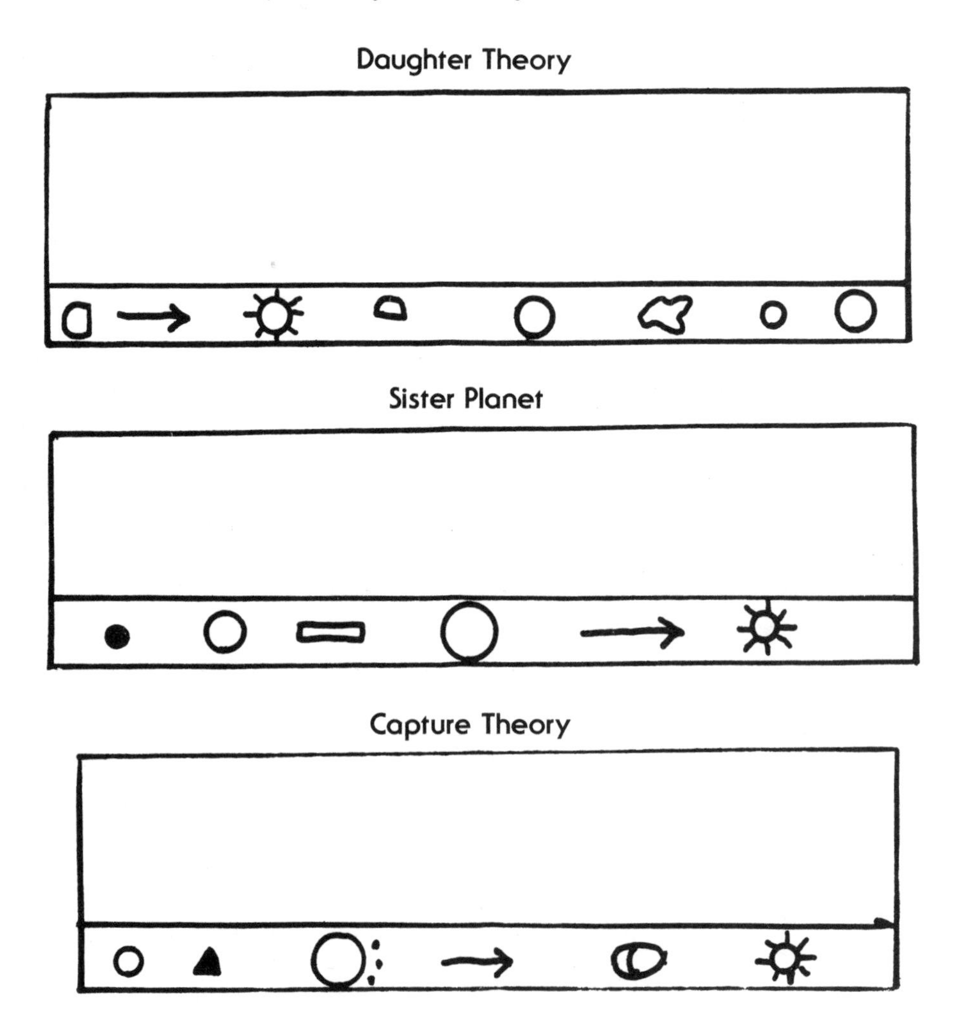

## PART 2

Answer these questions:

1. Which of the three theories seems more believable? Why?

________________________________________

________________________________________

________________________________________

________________________________________

2. List a weakness in each of the theories:

   a. Daughter theory: ________________________________________

   ________________________________________

   b. Sister planet theory: ________________________________________

   ________________________________________

   c. Capture theory: ________________________________________

   ________________________________________

3. If the moon was at one time part of the Earth, then what two things would hold true about moon rocks?

   a. ________________________________________

   b. ________________________________________

4. The bogus theory (not a real theory) says that about four billion years ago two asteroids—one, 4 miles in diameter and the other, 10 miles in diameter—collided and exploded. The Earth's gravity held the pieces together. In time, the moon was formed. List three problems with the bogus theory.

   a. ________________________________________

   ________________________________________

   b. ________________________________________

   ________________________________________

   c. ________________________________________

   ________________________________________

Name ______________________ Date ______________________

# 11-8 Constellations in a Minute!

**Introduction:** Constellations are apparent groups or clusters of stars resembling mythological figures. The ancients named constellations after animals and people. Some well-known constellations are The Big Dipper; The Great Bear; Orion, the Hunter; Leo, the Lion; Draco the Dragon. Imagination, of course, played a large part in naming these star groups.

**Objective:** To make model illustrations of make-believe constellations.

**Materials:** Drawing paper, black or blue construction paper, small stiff paintbrush, paint, newspaper, pencil, paperclip, and copy machine paper (see teacher).

**Procedure:** Select either Part 1 or Part 2. Then answer the questions in Part 3.



## PART 1

1. Cover your work area (flat surface) with newspaper.
2. Lay the construction paper over the newspaper.
3. Dip the paintbrush in the paint. Gently tap the brush against the paint jar to remove excess paint. *Note:* An old toothbrush works well.
4. Remove the brush from the jar, and run your thumb across the bristles as you move the brush back and forth over the paper. Avoid dropping large globs of paint.
5. Repeat until the entire paper receives an adequate coating.
6. Make several "star" charts using different color combinations.
7. Set charts aside to dry.
8. While the charts are drying, make up any constellation pattern that challenges the imagination by sketching figures on drawing paper—for example, Constellation Unicorn.

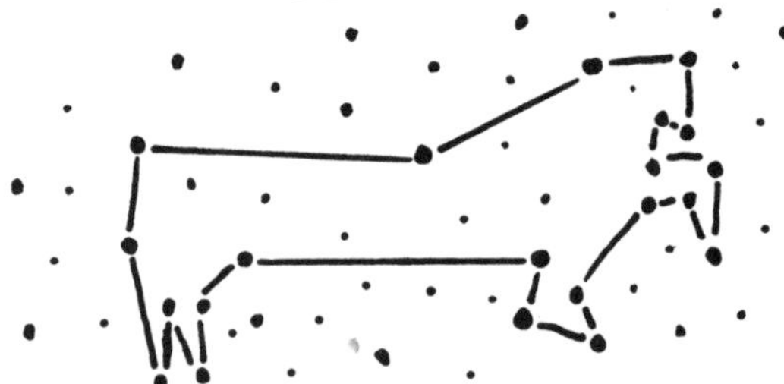

9. Now select a chart, locate a star pattern that fits your sketch, and reproduce the sketch by drawing it on the chart.

## PART 2

1. Draw several imaginary figures on paper. When you finish, pick out one that you wish to convert into a constellation.
2. Get one or two sheets of darkened copy machine paper from your teacher. This paper will represent the night sky.

3. Straighten out a paperclip. Use one end of the paperclip to scratch away the dark background on the paper. The scratching will produce white dots which resemble stars in the night sky.
4. Create a star pattern that matches your sketch. When you finish scratching, reproduce the sketch by drawing it on the chart.
5. Hold the chart in front of you. Tilt it slightly downward or to the side. Light reflecting off the paper helps illuminate the pattern.

## PART 3

Answer these questions:

1. If people today renamed the constellations, what patterns do you think might appear? Why?

______________________________

______________________________

______________________________

______________________________

2. How would a knowledge of constellation patterns help a person today?

______________________________

______________________________

______________________________

______________________________

Name ______________________ Date ______________________

# 11-9 Investigating a Mystery Planet

**Introduction:** You will investigate a newly discovered planet—Planet Mesmer. It represents a solid body, with different-colored lakes, and seems to have many geologic features similar to earth structures. How many items can you identify on Planet Mesmer?

**Objective:** To compare and contrast a miniature model of a "planet" with earth crustal features.

**Materials:** Observation Chart, test tube and tongs, burner, forceps, scoop, compound microscope and slides, magnifying glass, magnet, litmus paper, cobalt chloride paper (to test for the presence of water), beakers (50-100 ml), small jars (for collecting and separating materials), mortar and pestle, and thermometer (to take different soil level and liquid basin temperatures).

**Procedure:** Do the following:

1. Ask your teacher for an Observation Chart and model of Planet Mesmer.
2. Use all the available materials to thoroughly examine Planet Mesmer. Place all of your observations on the Observation Chart.
3. After you finish examining Planet Mesmer, use your findings to summarize the present condition of the mystery planet. Write a brief report, less than 200 words, about the discoveries you made. Use the back of this sheet if you need more space.

## Planet Mesmer Report

______________________________________________

______________________________________________

______________________________________________

______________________________________________

______________________________________________

______________________________________________

______________________________________________

______________________________________________

______________________________________________

______________________________________________

______________________________________________

______________________________________________

# Observation Chart

| Features | Basin No. 1 | Basin No. 2 | Geologic Structures | Organic Material | Other Observations |
|---|---|---|---|---|---|
| Composition | | | | | |
| Shape | | | | | |
| Color | | | | | |
| Location (N, E, S, W) | | | | | |
| Temperature | | | | | |
| How Like Earth? | | | | | |
| How Different from Earth | | | | | |
| Unique Features | | | | | |
| Present Condition | | | | | |
| Any Special Comment? | | | | | |
| Sketch or Drawing | | | | | |

Name ______________________ Date ______________________

# 11-10 Meteorite Impact

**Introduction:** Some scientists believe meteorites are fragments from the "missing planet" (a planet which some astronomers think once roamed between Mars and Jupiter). Other investigators believe meteorites come from the tails of comets. When meteorites approach the Earth, the gravitational field, like a powerful magnet, pulls them closer. The friction which builds up in the atmosphere creates intense heat.

A meteorite strikes the Earth, scatters, and sends shock waves racing in every direction. The depth of penetration and the crater size depend on where the meteorite hits the Earth (soft or hard terrain), its speed, and the size of the meteorite at impact. The crater is generally larger than the meteorite.

**Objective:** To simulate the impact of a meteorite.

**Materials:** Three different-sized steel bearings (small, medium, and large), clay, toothpick, metric ruler, cardboard (for protective covering).

**Procedure:** Do the following:

1. Make a clay pad 5 inches long, 3 inches wide, and 1 inch thick. The pad will represent the Earth's crust.
2. Ask your teacher for three steel bearings—small, medium, and large. Each bearing will represent a meteorite.
3. Place the clay pad in the center of the laboratory table or on the floor.
4. Stand over the clay pad and hold the bearing at shoulder level between your thumb and index finger. Center the bearing over the pad, and release it.
5. Drop each bearing, one at a time, on the clay pad.
6. After the first bearing hits the clay, remove it carefully. Then measure (mm) and record the depth and diameter of the crater on the recording chart.
7. Use a toothpick to measure the depth of the crater in the following manner: Hold the toothpick in a vertical position directly over the center of the crater. Lower the toothpick until it touches the bottom of the crater. Use a pen or pencil to mark the toothpick where it is level with the surface of the clay pad. Then measure the distance (mm) between the mark and the end of the toothpick that touched the bottom of the crater.
8. Repeat the procedure for the remaining bearings. Place all measurements on the recording chart.
9. Cover the laboratory table or floor with cardboard for protection. Set the clay pad on the cardboard. Make sure to smooth out the old crater marks.
10. Now forcefully toss the bearings into the clay pad by flipping them with a sharp downward snap of the wrist. This action, of course, will increase the speed of the falling bearings.
11. Measure and record the depth and diameter of the crater on the recording chart.

Answer these questions:

1. Which bearing makes the largest crater?

2. Which bearing travels the fastest? the slowest?

3. Are the crater and "meteorite" the same diameter? Explain.

4. What factors determine the speed of a meteorite?

5. Why haven't more meteorite craters been discovered?



| Soft Drop<br>METEORITE | DIAMETER OF IMPACT (mm) | DEPTH OF CRATER (mm) |
|---|---|---|
| Small | | |
| Medium | | |
| Large | | |

| Fast Drop<br>METEORITE | DIAMETER OF IMPACT (mm) | DEPTH OF CRATER (mm) |
|---|---|---|
| Small | | |
| Medium | | |
| Large | | |

# Section 11: Astronomy
## TEACHER'S GUIDE AND ANSWER KEY

### 11-1. Astronomy Word Search

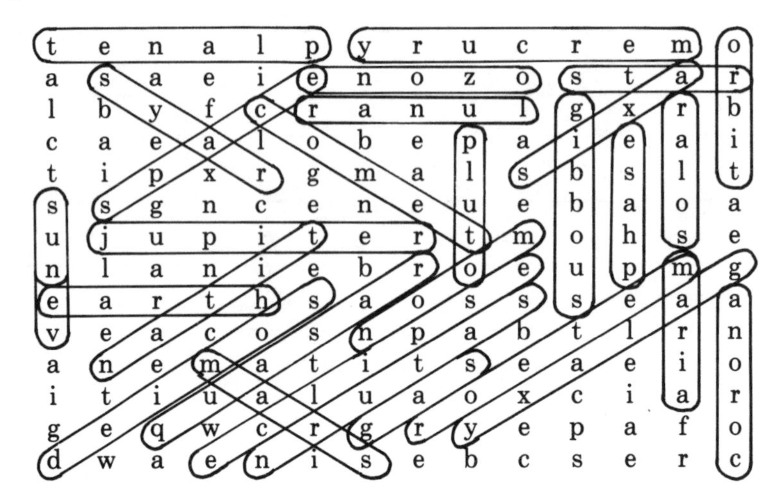

### 11-2. Moon Term Game

1. astronauts; 2. moondust; 3. Tycho; 4. crater; 5. perigee; 6. apogee; 7. orbit; 8. crescent; 9. gibbous; 10. full moon; 11. eclipse; 12. maria; 13. gravity; 14. first quarter; 15. phase

(**Answer to question:** It was *once part of Earth.*)

### 11-3 Nebular Theory

#### *Part 1*

1. planet; 2. nebula; 3. matter; 4. scientist; 5. originate; 6. central; 7. clouds; 8. gases; 9. space; 10. material; 11. escape; 12. shrank

#### *Part 2*

1. originate; 2. material; 3. matter; 4. planet; 5. clouds; 6. escape; 7. shrank; 8. central; 9. scientist; 10. space; 11. nebula; 12. gases

#### *Part 3*

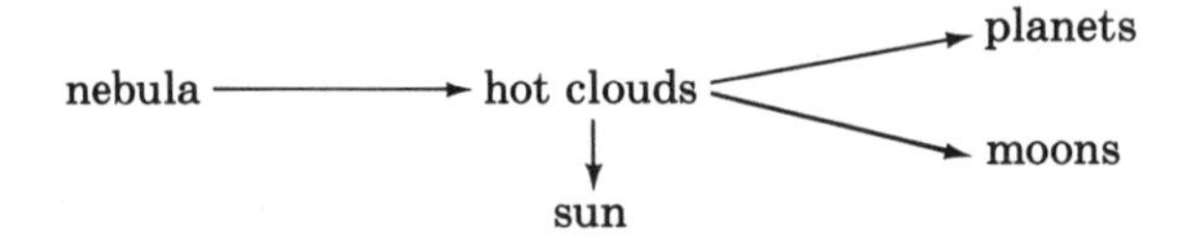


### 11-4 Planet Match-up

#### *Part 1*

Mercury—12; Venus—6; Earth—10, 5; Mars—7, 8, 11; Jupiter—2, 14; Saturn—1, 3; Uranus—1, 13; Neptune—4; Pluto—5, 9, 15

*Part 2*

1. Mercury; 2. Uranus; 3. Jupiter; 4. Mars; 5. Saturn

## 11-5 What Goes with What?

*Part 1*

1. h; 2. e; 3. l; 4. k; 5. j; 6. n; 7. b; 8. d; 9. a; 10. m; 11. c; 12. f; 13. i; 14. g

*Part 2*

1. Meteoroid—not a structure of the sun
2. Mare—a moon structure not related to the movement of the moon
3. Albedo—reflected light from the moon, not related to eclipse
4. Mercury—an inside, not an outside, planet
5. Cygnus—a galaxy, not a theory
6. Triangular—not one of the three classes of galaxies
7. Maria—a moon feature, not a star type
8. Red Spot—a feature of Jupiter, not Mars

## 11-6 How Do Planets Compare in Size (Diameter) with the Sun?

*Part 1*

Millimeter (mm) distances: Sun—86.6mm; Mercury—0.31 mm; Venus—0.77 mm; Earth—0.80 mm; Mars—0.42 mm; Jupiter—8.9 mm; Saturn—7.15 mm; Uranus—3.2 mm; Neptune—3.1 mm; Pluto—0.36 mm

*Part 2*

1. 25.36 mm, less than one-third; 2. slightly more than 9½; 3. slightly more than five times; 4. hours—2,165, days—90; 5. True statements will vary.

## 11-7 How About the Moon?

*Part 1*

Sketches will vary.

*Part 2*

**Possible answers:**

1. Answer will vary.
2. a. Too many differences in chemical makeup of the Earth and moon
   b. Not enough information
   c. Does not agree with the principles of gravity
3. a. Moon and Earth rocks would be alike chemically.
   b. Moon and Earth rocks would be the same age.
4. a. Chances of two asteroids colliding are very small.
   b. Asteroids are too small to make a large moon.
   c. "In time, the moon was formed" says little or nothing.

## 11-8 Constellations in a Minute!

You can get dark copy machine paper by running paper through the machine without setting something on the machine to be copied. White or yellow paper works well. Test different-colored paper to see what turns out the best. Have at least two sheets for each student. Remind students that they do not have to be artists to get excellent results.

*Part 3*

**Possible answers:**

1. Probably patterns resembling material objects or events of the day; e.g., space ships, athletes in action, high technology, automobiles, nuclear testing, and so on. These items would be newsworthy and interesting to most people.
2. It will help a person become familiar with the night sky, perhaps furthering his or her knowledge of ancient mythological folklore.

## 11-9 Investigating a Mystery Planet

Prepare several mystery planets ahead of time. Have on hand glass Pyrex™ or plastic trays to act as containers. You can prepare these planet models in several ways. Here is one suggestion:

1. Fill a large container (beaker or jar) with soil. Add water; mix, and pour the muddy batch into a Pyrex™ or plastic tray. Be sure to make enough to cover the entire tray (approximately 1 inch thick). One tray should be enough for four students.
2. Strengthen the mixture by adding small boulders, pebbles, or sand.
3. Mold valleys, canyons, hills, and mountain peaks.
4. Insert two small crucibles or beakers into the crust. These liquid-filled containers make excellent lakes or seas. Choose any of these aqueous recipes to fill the containers:
   a. Food coloring mixed with salt or sugar solution
   b. A diluted ammonia or alcohol bath
   c. Blue or green food coloring with plain water
   d. Pond water alive with microorganisms
   e. A yeast cell suspension (prepared by adding a half teaspoonful of powdered yeast to a bottle of sugar solution)
   f. Fruit punch or fruit-flavored carbonated soda
   g. White vinegar or apple cider vinegar
5. Sprinkle iron filings, charcoal powder, sulfur powder, or pieces of broken crystal (halite, calcite, or quartz) over the surface.
6. Poke small twigs, weeds, root fibers, or grass blades into the soil.
7. Allow the planet models to dry at room temperature for two or three days. The drying process produces cracks resembling fault lines, canyons, and valleys.

   The Observation Chart answers and Planet Mesmer reports will vary.

## 11-10 Meteorite Impact

1. The heaviest; 2. They travel at the same speed; 3. No, impact forms a larger crater size than meteorite; 4. Air friction, size and density of meteorite, etc.; 5. Many land in isolated areas or oceans, or craters erode away.

# Section 12: Astronomy
## MINI-ACTIVITIES

Listed below are 10 mini-activities, one to five minutes in length, for students to do at the beginning or end of the period.

1. The puzzle contains the letters that form the names of three planets. If you put the correct letters together, you'll use every letter in the puzzle only once.

| | | | |
|---|---|---|---|
| R | T | E | U |
| E | P | T | U |
| N | P | I | E |
| J | N | A | N |
| U | R | S | U |

**Planets**

1. ______________________

2. ______________________

3. ______________________

(**Answers:** 1. Neptune; 2. Uranus; 3. Jupiter

2. List two items (objects, names, etc.) well known in astronomy that have the letter combination "gal."

   (**Answer:** galaxy, galactic, Galileo, etc.)

3. Shade the puzzle spaces that contain the letters to the answer for the question: What is another name for asteroid?

   (**Answer:** planetoid)

   Unscramble the remaining letters in the puzzle to answer the question: Who developed the first reflecting telescope?

   (**Answer:** Newton)

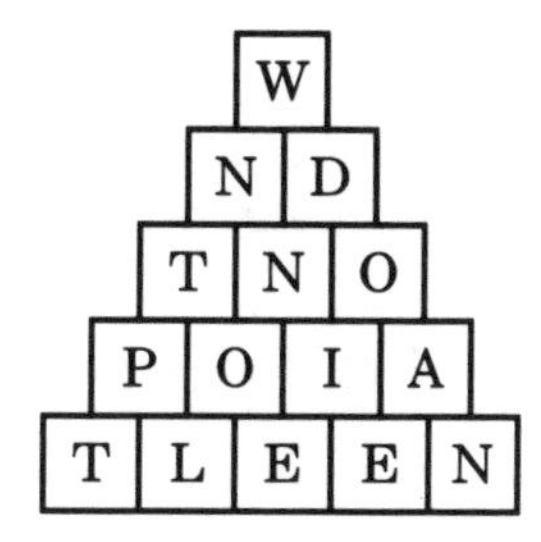


4. **Riddle:** Why does it stay so clean in outer space?

   (**Answer:** meteor showers)

5. Complete the puzzle.

**Across**

3. plural for spectrum
4. means "wandering"

**Down**

1. planet's diameter: 7,700 miles
2. means "seas"
5. a planet's orbital path

(**Answers:** *Across*—3. spectra; 4. planet; *Down*—1. Venus; 2. maria; 5. elliptical)

6. Each word group describes a space object. See if you can identify the objects.

   1. rock, metal, friction ______________________
   2. elliptical, spiral, irregular ______________________
   3. Polaris, luminous, celestial ______________________
   4. gas, corona, light ______________________

   (**Answers:** 1. meteor; 2. galaxies; 3. star; 4. sun)

7. List at least 10 items dealing with astronomy that begin with the letter "s."
   (**Possible answers:** Saturn, sun, solar, spectrum, spectral, spectroscope, satellite, sky, sphere, star, sunspot, etc.)

8. List at least 10 items dealing with any planet in the solar system that begin with the letter "e."
   (**Possible answers:** ellipse, Earth, ecliptic, erosion, equator, equinox, eccentric, eclipse, electromagnetic, element, etc.)

9 **Riddle:** What planet has an ear but cannot hear?
   (**Answer:** Earth)

10. **Question:** What do Earth, Mars, Saturn, and Uranus have in common?
   (**Answer:** They all have the letters "a" and "r" in their names.)

Section 12

# OCEANOGRAPHY

## OUTLINE

12-1 Oceanography Vocabulary Puzzle
12-2 Things in the Sea
12-3 Food Chain
12-4 Complete the Term
12-5 Secret Message
12-6 Instant Sea Water
12-7 Which Salt Solution Comes the Closest to 3.5%?
12-8 Depth Recording the Old-Fashioned Way
12-9 What Is the Mystery Object?
12-10 Miniature Model of a Density Current
12-11 Can You Beat El Niño?
12-12 Time to Dine

## MATERIAL

The following laboratory materials are needed for Section 12:

- apron
- balances
- beaker tongs
- beakers (50 ml, 100 ml)
- Erlenmeyer flasks (250 ml)
- food coloring
- game sheets (activities 12-11 and 12-12)
- glass plates
- graduated cylinders (100 ml)
- graph paper
- heat sources
- index cards
- markers (beans or popcorn kernels)
- matches
- metric rulers
- paper
- pencils
- pens
- prepared ocean basins (Activity 12-8)
- prepared salt solutions A, B, and C (Activity 12-7)
- safety goggles
- scissors
- scratch paper
- stirrers
- string or thread
- table salt
- timers (with second hands)
- washers or small weights
- water

Name ______________________ Date ______________

# 12-1 Oceanography Vocabulary Puzzle

Each clue refers to one of the words in the word list. Select the word that best fits each clue and write it in the space provided and in the puzzle. Be sure each choice exactly fits the puzzle.

## Across

1. Steep area at seaward edge of continental shelf: ______________
4. Steep mountain resting on ocean floor: ______________
6. Sea floor canyon or fissure: ______________
7. A section of water within the curve of a coastline: ______________
8. The up and down, back and forth movement of water across the surface: ______________
9. The great body of salt water that covers about two-thirds of the Earth's surface: ______________
10. Deep, flat areas of the ocean floor: ______________
13. A sonar signal sent through water: ______________
14. A fracture or break in sea floor rocks: ______________
16. A well-known undersea explorer: ______________
18. An undersea volcanic structure characterized by a broad, flat top: ______________
19. The regular rise and fall of ocean water: ______________
20. A small lump of elements; for example, manganese: ______________
21. Single-celled water plant with a silica shell: ______________

## Down

2. The study of the oceans: ______________
3. Floating brown seaweed found in the Sargasso Sea: ______________
5. The upward movement of sea water to the surface: ______________
6. A zone of rapid temperature change usually found near the surface of bodies of water: ______________

9. Deep-sea sediments containing large amounts of organic remains: ______________
11. The amount of dissolved solids in sea water; that is, how much salt is present: ______
______________________________________________
12. Location measured in degrees north and south of the equator: ______________
15. Minute floating organisms in surface waters: ______________
16. An effect of Earth's rotation that causes the deflection of objects which are moving toward or away from the equator: ______________
17. An undersea structure that extends from a lower to higher elevation: ______________
______________________________________________

## Word List

| | | | |
|---|---|---|---|
| abyssal | guyot | plankton | sounding |
| coriolis | latitude | rise | thermocline |
| Cousteau | nodule | salinity | tide |
| diatom | ocean | sargassum | trench |
| fault | oceanography | seamount | upwelling |
| gulf | oozes | slope | waves |

Name ______________________ Date ______________________

# 12-2 Things in the Sea

Find these eight sea-related words hidden in the puzzle. Answers may be found vertically, horizontally, backward, or forward. Shade the spaces in the puzzle for all eight words. If you complete the puzzle correctly, the shaded spaces will spell the answer to this question: *What is located in the Arctic Sea that averages a thickness of 20 to 23 feet?*

## Word List

| | | | |
|---|---|---|---|
| baleen | canyon | seamount | slope |
| basin | ridge | shelf | swells |

| | | | | | | | | | | | | | | | | | |
|---|---|---|---|---|---|---|---|---|---|---|---|---|---|---|---|---|---|
| A | F | J | W | I | V | E | U | D | T | C | U | X | I | T | A | W | E |
| G | S | Y | G | B | F | C | N | K | V | M | E | J | F | K | S | A | B |
| X | E | T | F | H | C | F | D | R | L | A | G | I | O | A | X | Z | A |
| W | A | E | G | U | Z | G | H | Z | H | Y | N | L | T | B | Y | S | V |
| D | M | F | I | E | G | D | I | R | M | E | B | A | L | E | E | N | I |
| S | O | Q | H | P | R | J | Q | I | H | S | A | P | S | Q | W | T | P |
| V | U | M | S | O | V | O | P | Q | I | J | S | W | E | L | L | S | B |
| C | N | T | P | L | N | B | B | D | C | I | I | B | M | O | A | L | A |
| U | T | K | A | S | H | E | L | F | W | X | N | O | Y | N | A | C | Y |
| K | L | R | O | A | J | P | D | O | R | Y | Z | L | A | S | I | T | A |



*Just for Fun:* Unscramble the following letters and find out the mystery sea animal (two words).

A R K I N G H B A S S K

Answer: ______________________

Name ______________________ Date ______________________

# 12-3 Food Chain

The sea teems with plant and animal life. Researchers estimate that 90% of the world's marine life can be found within 200 miles of land.

Microscopic algae known as diatoms belong to the plankton family. They are one-celled plants with a silica covering. They supply sea animals with an abundance of food. The nekton layer that lies beneath the plankton zone includes the swimming marine organisms. These animals swim about, change depths, and feed on varieties of plankton. Nekton organisms include adult fish—tuna, bass, bonito, salmon—squid, sea horses, sea turtles, seals, and whales. The benthos layer contains bottom-dwelling organisms such as algae, sea worms, anemones, crabs, starfish, sponges, clams, and snails.

Many food chains exist in the sea. Plants begin the chain by providing food for other organisms. The plant-eating organism, in turn, becomes a meal for a larger animal. Plants convert sunlight, carbon dioxide, and water into proteins, carbohydrates, and starches by a process called photosynthesis. Therefore, when an animal consumes a plant organism, food energy is passed along to the larger animal. What if the larger animal is eaten by still a larger organism? Again, the food energy moves from one organism to another.

Now do the following:

1. Select 10 different organisms from the Marine Life Chart. In the space below show, using arrows, how the flow of energy moves from one organism to another in a food chain.



2. Repeat the procedure. This time select only three organisms from the chart.

# Marine Life Chart

| Zone | Organism | Eats | Eaten By |
|---|---|---|---|
| Plankton | Diatom (yellow-green algae) | Ingests or absorbs complex organic molecules | Protozoa, fish, clams, snails, copepods, foraminifera, etc. |
| Plankton | Dinoflagellate | Ingests or absorbs complex organic molecules | Protozoa, fish, clams, snails, copepods, foraminifera, etc. |
| Plankton | Radiolarian | Diatoms and dinoflagellates | Mackeral, herring, anchovies, whales, sardines |
| Plankton | Foraminifera | Diatoms and dinoflagellates | Marine animals; same species as next step above |
| Nekton | Squid | Crustaceans, mollusks, fish | Marine mammals, fish, humans |
| Nekton | Seal | Fish | Large marine mammals, sharks |
| Nekton | Dolphin | Fish | Large marine mammals |
| Nekton | Tuna | Fish | Humans, sharks, large marine mammals |
| Benthos | Chiton | Microorganisms, seaweed | Humans (West Indies) |
| Benthos | Starfish | Clams, oysters, mollusks, worms | Dolphins |
| Benthos | Clam | Microorganisms, diatoms | Starfish, fish, humans |
| Benthos | Sponge | Microorganisms, organic matter | Seldom eaten by other sea creatures |

Name ________________________________ Date ____________________

# 12-4 Complete the Term

Complete each oceanography term by unscrambling the letters in the right-hand column and writing them in the appropriate spaces next to the incomplete term in the left-hand column. Also, write the letters—a, b, c, etc.—in the spaces at the left. Use the hints in parentheses to help you complete each term.

| Term | Hint | |
|---|---|---|
| __ 1. ben ______ | (water zone) | a. ntok |
| __ 2. sali ______ | (salt) | b. main |
| __ 3. hydro ______ | (Earth's waters) | c. fr |
| __ 4. gu ______ | (flat-topped) | d. lacti |
| __ 5. thermo ______ | (temperature change) | e. gdni |
| __ 6. cur ______ | (moving water) | f. mtoh |
| __ 7. coast ______ | (water touches land) | g. roaoint |
| __ 8. at ______ | (coral reef) | h. snu |
| __ 9. sand ______ | (sedimentary rock) | i. ynti |
| __ 10. no ______ | (lump) | j. yto |
| __ 11. evap ______ | (liquid to gas) | k. einl |
| __ 12. soun ______ | (sonar signal) | l. oths |
| __ 13. fa ______ | (equals 6 feet) | m. noset |
| __ 14. sl ______ | (beyond the shelf) | n. eshper |
| __ 15. mi ______ | (very low tide) | o. epo |
| __ 16. prec ______ | (moisture—i.e., rain) | p. tner |
| __ 17. nau ______ | (ships or navigation) | q. lol |
| __ 18. su ______ | (breaker zone) | r. elcin |
| __ 19. tsu ______ | ("tidal zone") | s. eldu |
| __ 20. plan ______ | (floating organisms) | t. piatanoit |

Name ______________________ Date ______________________

# 12-5 Secret Message

Fill in the correct word for each clue below. Then place the numbered letters in the correct blanks of the message at the bottom of the page to find the secret message.

| Word blanks | Clue |
|---|---|
| _ _ _ _ _ _ _ _ _ _ (20, 35, 28, 5, 6) | Materials deposited by water |
| _ _ _ _ _ _ (4, 24) | Associated with the sea; maritime |
| _ _ _ _ _ _ _ _ _ (23, 7, 30) | Living plants or animals |
| _ _ _ _ _ _ _ (34, 1, 29) | Organisms living on the sea bottom |
| _ _ _ _ _ _ (25) | Organisms swimming on or near the surface of the sea |
| _ _ _ _ _ _ (40) | One-celled plant with silica shell |
| _ _ _ _ (37) | One nautical mile per hour |
| _ _ _ _ _ _ _ _ _ _ _ _ _ _ _ _ (two words) (27, 2, 8, 13, 22) | Currents caused by sediment-laden water |
| _ _ _ _ _ _ _ _ (two words) (3, 10, 19, 33, 9, 32) | The level of the surface of the ocean between low and high water |
| _ _ _ _ _ _ _ (41, 14, 39) | A giant sea wave generated by an earthquake or sea bottom disturbance |
| _ _ _ _ _ (31, 38, 18) | Ring-shaped coral reef in deep ocean |
| _ _ _ _ _ _ _ _ _ _ (15, 17) | The physical features of the ocean bottom |
| _ _ _ _ _ _ _ (16, 12, 11, 36) | Samples of sediments taken from the ocean floor |
| _ _ _ _ _ _ _ _ _ _ (two words) (21, 26) | Another name for sea bottom |

***Secret Message:***

_ _ _ _ _ _ (1 2 3 4 5 6) _ _ _ (7 8 9) _ (10) _ _ _ _ _ (11 12 13 14 15)

_ _ _ _ _ _ _ _ _ (16 17 18 19 20 21 22 23 24) _ _ (25 26) _ _ _ _ _ _ _ _ _ (27 28 29 30 31 32 33 34 35)

_ _ _ _ _ _. (36 37 38 39 40 41)

Name ______________________________ Date ______________________

# 12-6 Instant Sea Water

**Introduction:** The Earth's rivers dump millions of tons of dissolved salts and other materials into the oceans each year. Runoff of rain from the land carries these materials to the rivers which, in turn, wash into the oceans. The water dissolves the salts, thus making the oceans salty. One thousand grams of sea water contain, on the average, 35 grams of dissolved salts. Therefore, dissolved salts account for 3.5% of the weight of sea water.

**Objective:** To make "sea water" with a 3.5% salinity.

**Materials:** Beakers (50, 100 ml), beaker tongs, water, stirrers, heat source, balances, table salt, pencils and paper, safety goggles, matches, and graduated cylinders (100 ml).

**Procedure:** Weigh out 7 grams of salt. Add the salt to 200 ml of water. Mix thoroughly. This will be the prepared salt solution.



## PART 1

Now determine the percentage of salt by weight in the following manner:

1. Weigh a dry beaker. __________g
2. Measure 10 ml of prepared salt solution in a graduated cylinder. Pour the solution into a beaker. Weigh the solution and the beaker. __________g
3. Determine the weight of the solution. __________g
4. Evaporate the solution by heating it over low flame. Do not bring the solution to a vigorous boil. (Be sure to wear safety goggles.) When the solution evaporates, remove the beaker from the heat source with beaker tongs. Set the beaker aside and allow it to cool.
5. Weigh the beaker and residue (solid material in beaker). __________g
6. Determine the weight of the residue. Subtract the dry beaker weight (Step 1) from the weight of the beaker and residue (Step 5). __________g
7. Calculate the percentage of salt. Divide the weight of the solution (Step 3) into weight of the residue (Step 6). Multiply the answer by 100. __________% salt

## PART 2

Answer the following questions on the back of this page:

1. What do you think would eventually happen to sea life if the rivers suddenly dried up?
2. What do you think would eventually happen to sea life if the salt concentration of the water rose from 3.5% to 8% or 9%?
3. What factors may have led to a higher or lower salinity concentration (3.5%) in the experiment?

Name ______________________ Date ______________________

# 12-7 Which Salt Solution Comes the Closest to 3.5%?

**Introduction:** You will receive three salt solutions with different salinity concentrations. One solution comes the closest to 3.5%. Which one? You'll have to test each solution to find out.

**Objective:** To test three different salt solutions to discover which one is closest to 3.5% salinity.

**Materials:** Beaker (50 ml), beaker tongs, Solutions A, B, and C, stirrers, heat source, balances, safety goggles, graduated cylinders (100 ml), matches, and pencils and paper.

**Procedure:**

## PART 1

Follow the procedure for determining the percentage of salt by the weight given in Activity 12-6: Instant Sea Water. Record your answers in the chart below:

| Solution | Wt. of Water | Wt. of Beaker and Solution | Wt. of Solution | Wt. of Beaker and Residue | Wt. of Residue | % Salt |
|---|---|---|---|---|---|---|
| A | | | | | | |
| B | | | | | | |
| C | | | | | | |

Solution ______ comes closest to 3.5%.

## PART 2

Answer the following questions on the back of this page:

1. Joan ran into a problem. She found that the weight of the beaker and residue measured less than the weight of the dry beaker. How could this be?
2. If you were Joan, what steps would you take to correct the error?
3. Fred's teacher gave him 500 ml of 10% salt solution and asked him to find two ways to decrease the salinity of the solution. If you were Fred, what two things would you do?

Name ______________________ Date ______________________

# 12-8 Depth Recording the Old-Fashioned Way

**Introduction:** Today, scientists use sonar, a detection system based on the reflection of underwater sound waves, to map the ocean floor. They can determine the depth and shape of objects lying on the ocean bottom with great precision. Sound waves travel about 4,700 feet per second through water. From this figure, water depth can easily be calculated. The farther away an object is, the longer it takes the sound wave to reach the receiver. The total sounding time divided in half gives the depth of the object under water.

**Objective:** To simulate how soundings record the depth and shape of objects lying on the ocean bottom.

**Materials:** Prepared ocean basins, string or thread, washers or small lead weights, metric rulers, graph paper, and pencils and paper.

**Procedure:** Obtain a model ocean basin from your teacher. Tie string or thread to a washer or fishing weight. Make sure the string/weight combination is long enough to reach the bottom of the tray.

Go to the first mark on the tray. Lower the weight into the water. After it hits bottom, mark the area where the surface of the water meets the string by pinching the string with your thumb and index finger. Lift the weight from the tray. Measure the string length with a metric ruler. Record the measurement in centimeters on paper. (Assume 1 centimeter equals 100 feet.)

Repeat this procedure until measurements are taken for all marks.

Graph your results by placing *ocean depth* on the vertical axis and *marked number* on the horizontal axis.

Briefly describe the contour of the model ocean basin from the graph results.



---

Answer the following questions on the back of this page:

1. Let's suppose you were taking a sonar profile of a deep ocean area. You send out a signal, but it doesn't return. You release three more signals. None return. How would you account for the missing signals?
2. Many fishing boat captains use sonar equipment on their vessels. What are three reasons for using such equipment?

Name ______________________ Date ______________________

# 12-9 What Is the Mystery Object?

**Introduction:** Sonar readings taken at different times over the same general area are called profiles. Profiles reveal such ocean floor features as plains, deep trenches, plateaus, hilltops, and mountain ranges.

**Objective:** To see how sounding data reveals the depth and shape of objects lying on the ocean bottom.

**Materials:** Graph paper, rulers, and pencils and paper.

**Procedure:** Do the following:

1. Construct a vertical axis and horizontal axis on graph paper. Label the vertical axis *ocean depth* (feet). Use a range of 1,350 to 1,420 feet. Place *distance ship traveled* (feet) on the horizontal axis. Let each line along the horizontal axis equal 5 feet.
2. Now plot the data from the chart on graph paper.

### Mystery Object Data Chart

| Distance Ship Traveled (ft) | Ocean Depth |
|---|---|
| First Signal Reported | 1400 |
| 25 | 1400 |
| 30 | 1380 |
| 35 | 1380 |
| 40 | 1380 |
| 45 | 1365 |
| 50 | 1365 |
| 55 | 1400 |
| 60 | 1400 |
| 65 | 1400 |
| 70 | 1410 |
| 75 | 1410 |
| 80 | 1410 |

3. Make a profile by drawing a line from dot to dot. Describe what you think the profile reveals. ______________________

Answer the following questions on the back of this page:

1. Do you think scientists could have found the Titanic without sonar equipment? Why or why not?
2. A ship sends out a sonar signal. Between the time the signal left and when it returned, 2.6 seconds elapsed. If sound waves travel about 4,700 feet per second through water, what was the ocean depth? *Hint:* The total sounding time divided in half gives the depth of the object under water.

Name ______________________ Date ______________________

# 12-10 Miniature Model of a Density Current

**Introduction:** Simply stated, cold water sinks, warm water rises. Strong density currents that flow deep beneath the surface are caused by the sinking of water. Cooling and an increase in salt content add weight to the water, thus causing water to sink. For example, cold water near the poles sinks, and warm water near the equator rises. The cold water sinks and moves toward the equator.

**Objective:** To observe how cold, heavy water sinks and warm water rises.

**Materials:** Erlenmeyer flasks (250 ml), glass plates, apron, cold and warm water, beakers (250 ml), food coloring, salt, and pencils and paper.

**Procedure:** Do the following:

## PART 1

1. Fill one flask with warm water. Add two drops of food coloring. Place your hand over the mouth of the flask and shake it. Fill a second flask with cold water.
2. Place a glass plate over the mouth of the flask of cold water, turn it over, and set it on top of the flask containing warm, colored water. Hold both flasks in an upright position making sure the mouths of the flasks line up properly. Remove the glass plate.
3. Record what happens to the warm, colored water. ______________________

   ______________________
4. Describe how the results may simulate the action of density currents. ______________________

   ______________________
5. Empty and rinse the flasks.

## PART 2

1. Cover the bottom of a flask with salt. Fill the flask with water. Add two drops of food coloring. Place your hand over the flask and shake vigorously. Dissolve as much salt as possible. Fill a second flask with clear, fresh water.
2. Place a glass plate over the mouth of the flask of salty, colored water, turn it over, and set it on top of the flask containing clear, fresh water. Hold both flasks in an upright position making sure the mouths of the flasks line up properly.
3. Remove the glass plate.
4. Record what happens to the salty, colored water. ______________________

   ______________________

5. Describe how the results may simulate the action of density currents. ______________

______________________________________________

6. Empty and rinse the flakes.

## PART 3

Answer the following questions:

1. What natural event might increase salinity and cause density currents to increase speed in certain areas?

______________________________________________

______________________________________________

______________________________________________

______________________________________________

______________________________________________

______________________________________________

2. Does this activity prove how density currents develop? What does it show?

______________________________________________

______________________________________________

______________________________________________

______________________________________________

______________________________________________

______________________________________________

3. If fresh water replaced salt water, what might happen to ocean currents as we now know them?

______________________________________________

______________________________________________

______________________________________________

______________________________________________

______________________________________________

______________________________________________

Name ______________________ Date ______________________

# 12-11 Can You Beat El Niño?

**Introduction:** Ocean currents, to some degree, affect weather patterns. El Niño, for example, is a warm current of equatorial water that usually appears off the coast of South America around Christmas. It arrives in cycles, causing extreme heatings of ocean temperatures every four or five years.

In the spring of 1982, El Niño's appearance marked significant changes. It brought a reversal in the Western Pacific trade winds. These winds normally blow warm surface waters westward, away from the Americas. However, in 1982, the trade winds weakened, blew in reverse, and sent warm water eastward toward the Americas.

The temperature and wind shifts created worldwide problems. Droughts ravaged Australia, Indonesia, and parts of Africa; heavy storms pelted California beaches; and the commercial fish catch in South America declined sharply.

The increased water temperature in some areas killed fish and other forms of sea life. Marine organisms migrated to more desirable habitats. This sudden change, of course, upset the food cycle and created mass confusion.

Was El Niño totally destructive? No. In fact, researchers say El Niño created unique conditions favoring many forms of life. Some fish and shellfish species increased their range and abundance off the western coast of South America. In addition, scientists have reported rainfall to arid regions, which allowed people to farm and grow crops. Many marine species flourished during this time.

**Objective:** To understand how ocean currents can influence weather conditions, environments, and economic conditions.

**Materials:** Index cards, El Niño game chart (see teacher), paper, and pencils and pens.

**Procedure:** Do the following:

1. Copy each of the numbered items from the El Niño game chart onto a separate index card. Example:

   Item 1

   Increased vegetation

   +1 point value

2. Number 1 through 10 on a piece of paper.
3. Shuffle the index cards and place them face down on a desk or table.
4. Remove the first card from the pile, read the item, and write its point value opposite number 1 on the paper. Set the card in a separate stack away from the pile. Do not reshuffle the cards.
5. Draw a second card from the pile, read the item, and write its point value opposite number 2 on the paper. Repeat this procedure until you draw 10 cards and record their point values.
6. Total the point values. For example, if your scores read: 1. +1; 2. −2; 3. 0; 4. −½; 5. −4; 6. +3; 7. −2; 8. −1½; 9. +2½; 10. 0, your total score equals −3½.

7. If you record a + or positive point value, you beat El Niño. Any – or negative value score means El Niño wins, you lose.
8. Play "Best of Five." The first to win three games—you or El Niño—becomes champion.

At the end of the game, take the cards you end up with and write a brief paragraph based on the cards' information.

## El Niño Game Chart

| Number | Item | Point Value |
|---|---|---|
| 1 | Increased vegetation | +1 |
| 2 | Increased numbers of fish/shellfish | +3 |
| 3 | Newly discovered organisms | +2 |
| 4 | Marine organisms flourish | +2 1/2 |
| 5 | Arid regions get rain | +2 |
| 6 | Wider range of fish/shellfish | +2 1/2 |
| 7 | Drought | –2 |
| 8 | Hurricane | – 1/2 |
| 9 | Economic crises | –4 |
| 10 | Many dead fish | –3 |
| 11 | Fish migrate; leave area | –1 1/2 |
| 12 | Broken food chain | –2 |
| 13 | You draw a blank | 0 |
| 14 | You draw a blank | 0 |

Name ______________________ Date ______________________

# 12-12 Time to Dine

**Introduction:** The food chain is a complex series of events. Living things are directly linked to one another by what they eat. Undoubtedly, the food chain is a complicated mesh of interconnecting branches. One set pattern of food energy does not exist. For instance, plankton may be eaten by a crustacean known as a copepod, protozoa, clams, fish living in the nekton zone, sponges from the benthos zone, basking sharks, and so on.

**Objective:** To demonstrate how the food chain is a complex series of events.

**Materials:** 5 × 7-inch index cards, rulers, pencils or pens, scissors, and paper.

**Procedure:** Do the following:

1. Cut out a 5″ diameter circle from an index card.
2. Mark off the card in eight equal sections.
3. Write these words on the cardboard wheel, one word per section: *diatom, foraminifera, copepod, sardine, tuna, shark, humans,* and *oil spill. Note:* These organisms represent an energy flow in the food chain. They appear as one continuous branch in a chain.
4. Push a pencil or pen through the center of the wheel. Rotate the pencil to widen the hole. The wheel should spin freely around the pen or pencil.
5. Objective: To turn the wheel until all organisms in the food chain appear.

Here are the rules for playing "Time to Dine":

1. Choose a partner. Each person takes a turn spinning the wheel.
2. The player starts the activity by spinning the wheel. The spinning player's thumbnail acts as a pointer. When the wheel comes to a stop, the spinner calls out the section resting nearest to his thumb.

   The spinner doesn't write the name of an organism on paper until it appears in its proper order—e.g., first, *diatom;* second, *foraminifera;* third, *copepod,* and so on.
3. If the wheel stops between sections, the player spins again. The spinning player writes down the name of each section only once.
4. Each time *oil spill* appears, the spinning player must cross out his or her last selection. For instance, the player has listed *diatom, foraminifera,* and *copepod.* He spins *oil spill.* He must cross out *copepod.* The player cannot advance until he spins *copepod* again.
5. Continue taking turns spinning the wheel. The player who completes his or her list first wins the game.

At the end of the game, use what you have written down as a result of playing this game and write a paragraph about what you have learned.

# Section 12: Oceanography
# TEACHER'S GUIDE AND ANSWER KEY

## 12-1 Oceanography Vocabulary Puzzle

*Across:* 1. slope; 4. seamount; 6. trench; 7. gulf; 8. waves; 9. ocean; 10. abyssal; 13. sounding; 14. fault; 16. Cousteau; 18. guyot; 19. tide; 20. nodule; 21. diatom *Down:* 2. oceanography; 3. sargassum; 5. upwelling; 6. thermocline; 9. oozes; 11. salinity; 12. latitude; 15. plankton; 16. coriolis; 17. rise

## 12-2 Things in the Sea

| A | F | J | W | I | V | E | U | D | T | C | U | X | I | T | A | W | E |
|---|---|---|---|---|---|---|---|---|---|---|---|---|---|---|---|---|---|
| G | S | Y | G | B | F | C | N | K | V | M | E | J | F | K | S | A | B |
| X | E | T | F | H | C | F | D | R | L | A | G | I | O | A | X | Z | A |
| W | A | E | G | U | Z | G | H | Z | H | Y | N | L | T | B | Y | S | V |
| D | M | F | I | E | G | D | I | R | M | E | B | A | L | E | E | N | I |
| S | O | Q | H | P | R | J | Q | I | H | S | A | P | S | Q | W | T | P |
| V | U | M | S | O | V | O | P | Q | I | J | S | W | E | L | L | S | B |
| C | N | T | P | L | N | B | B | D | C | I | I | B | M | O | A | L | A |
| U | T | K | A | S | H | E | L | F | W | X | N | O | Y | N | A | C | Y |
| K | L | R | O | A | J | P | D | O | R | Y | Z | L | A | S | I | T | A |

*Mystery sea animal:* basking shark

## 12-3 Food Chain

Answers will vary.

## 12-4 Complete the Term

1. benthos, l; 2. salinity, i; 3. hydrosphere, n; 4. guyot, j; 5. thermocline, r; 6. current, p; 7. coastline, k; 8. atoll, q; 9. sandstone, m; 10. nodule, s; 11. evaporation, g; 12. sounding, e; 13. fathom, f; 14. slope, o; 15. minus, h; 16. precipitation, t; 17. nautical, d; 18. surf, c; 19. tsunami, b; 20. plankton, a

## 12-5 Secret Message

sediments, marine, organisms, benthos, nekton, diatom, knot, turbidity current, sea level, tsunami, atoll, topography, corings, ocean floor

*Secret message:* Oceans are a great collector of dissolved solids.

## 12-6 Instant Sea Water

Remind students to weigh materials carefully and to wear safety goggles while heating solutions.

### *Part 2*

1. Many organisms would die; some would survive and adjust to a new environment.
2. Most likely the answer to Question 1 would also apply here, depending on how fast the change in salt concentration occurred.
3. Errors in measurement, adding too much salt to the water, errors in calculation, weighed beaker and residue before completely dry, etc.

## 12-7 Which Salt Solution Comes the Closest to 3.5%?

Remind students to weigh materials carefully and to wear safety goggles while heating solutions. This activity will take more than one class period to complete.

### *Part 1*

Take three large containers—A, B, and C—and pour 1,000 ml of water into each one. Add 70 grams of sodium chloride to beaker A, 30 grams to beaker B, and 100 grams to beaker C. Stir thoroughly. Prepare these solutions before assigning the activity to students. Adding the food coloring to each solution heightens student interest. The approximate percentage of each solution is: A—7%, B—3%, and C—10%. Solution B should come the closest to 3.5%.

### *Part 2*

1. She may have used two different beakers, misread the balance, made errors in calculation, etc.
2. Redo the experiment, make careful measurements, and keep track of laboratory equipment.
3. Add fresh water to the solution, let salt settle to the bottom of the container and slowly pour surface water into a second container, etc.

## 12-8 Depth Recording the Old-Fashioned Way

Prepare several model ocean basins in advance. Follow these steps:

1. Place various-sized rocks and pebbles along the bottom of a tray.
2. Fill a large container with water. Add several drops of food coloring. Stir the contents. *Note:* The food coloring prevents students from seeing the bottom of the tray.
3. Pour the colored water into the tray until nearly full.
4. Lay a strip of masking tape along a side edge of the tray. Make marks along the tape, one mark every 2 centimeters, until you cover the length of the tray. Number the marks from left to right.
5. Set the trays aside for the students to use.

Use several examples to show students how to prepare a line graph.
Contour descriptions will vary with each basin.

### *Answers to Questions:*

1. Perhaps the receiver is broken, perhaps the sending unit isn't functioning properly, possibly the signal lost energy by traveling so far, etc.
2. Locating fish, pinpointing their depth, and mapping the contour of the bottom

## 12-9 What Is the Mystery Object?

If necessary, review with students how to construct a line graph. The mystery object or profile could be a sunken ship, a rock formation, or an unknown structure.

1. Maybe, but it would have taken considerable luck and time. The navy submersibles with their video cameras needed sonar help to uncover likely areas where the Titanic might be resting.
2. $4{,}700 \times 2.6 = 12{,}220$ feet. $12{,}220 \div 2 = 6{,}110$ feet. Answer: 6,110 feet

## 12-10 Miniature Model of a Density Current

Diagram on the board, using arrows, how cold, heavy water sinks and warm water rises. Prepare cold and warm water in advance.

### *Part 1*

The heavier, cold water sinks into the lighter, warm water causing it to circulate or move about. The results simulate the action of density currents by showing how heavier water sinks causing circulation to speed up.

### *Part 2*

Students should see that the heavier salt solution speeds up water circulation. The results simulate the action of density currents by showing how heavier water sinks.

### *Part 3*

1. Evaporation of sea water in shallow areas increases salt content, thus making the remaining water heavier. This phenomenon occurs at Gibraltar, where the Mediterranean Sea and Atlantic Ocean join.
2. This activity is a model or simulation showing a simple concept—warm water rises and cold water sinks. It does not attempt to prove anything.
3. Water would still rise and fall from the gravitational attraction between the sun, moon, and Earth. Some ocean currents would probably weaken and slow down.

## 12-11 Can You Beat El Niño?

Discuss El Niño with students. Use a wall chart or world globe to show students the location of El Niño.

## 12-12 Time to Dine

List four or five organisms on the chalkboard. Have students help you devise a food chain using arrows to show the direction of energy flow. For example:

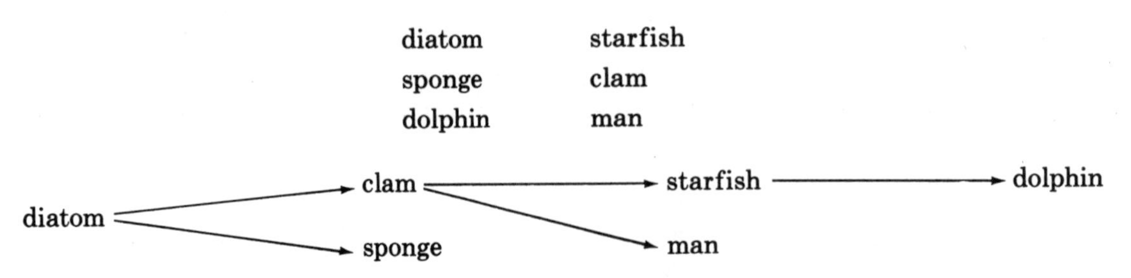

# Section 12: Oceanography
## MINI-ACTIVITIES

Listed below are 14 mini-activities, one to five minutes in length, for students to do at the beginning or end of the period.

1. Complete the following limerick. You may use no more than nine syllables in the last line.

   There once was a lady named Dee,
   Whose job was to study the sea;
   She loved to chase dolphins,
   Net sardines and sculpins,

   ____________________________.

   (Answers will vary.)

2. 1,500 grams of water contains 7.5 grams of dissolved salts. Calculate the percentage of salinity. __________% salinity (Divide 1,500 grams into 7.5 grams. Multiply the answer by 100. Answer = 0.5%)

3. How about another limerick? Remember, use no more than nine syllables in the last line.

   There once was a guy named Del,
   Whose job was to dive in a bell;
   One day while exploring,
   His sleep brought on snoring,

   ____________________________.

   (Answers will vary.)

4. Fill in the blanks with the correct letters. Put the circled letters together to answer the question below.
   1. Science of ocean study: _ c _ (_) _ (o) _ r _ _ _ _ _
   2. A steep undersea canyon: (_) _ _ (n) _ h
   3. Continuous movement of ocean water: _ (u) _ _ (_) n _
   4. One-celled marine plants: _ _ a _ _ (m) (_)

   What is an undersea, steep-sided, mountainlike structure called?

   _ _ _ _ _ _ _ _

   (**Answers:** 1. oceanography; 2. trench; 3. current; 4. diatoms; Answer to the question is *seamount.*)

5. Complete the following statements with a word related to the study of the sea. For example,

   *"Sorry, I can't help you. My hands are ________________." (tide)*

   Corny or not (or should I say knot?), here we go:

   1. A recent happening is a ________________ event. (current)
   2. A pirate might say, "You'll have to walk the ________________, Tom." (plank)

3. "________________ goodbye to Aunt Mary, Doris." (wave)
4. My sister, Louise, will soon give birth. She doesn't care if the baby is a ________________ or ________________. (gull, buoy)
5. "Cheer up, Alex. ________________ or later things will get better." (sonar)
6. A message in a bottle washed ashore. Translate the message from the symbol clues.

   Sand + $1's "R" + + Mun +

   (Message: Sand dollars are not money.)
7. **Riddle:** Why did the cruise ship passenger fill his mattress with ocean current water? (**Answer:** He thought it would help him drift off to sleep.)
8. The figures are related to life or events in the sea. What are they? (*Hint:* Use your imagination.)

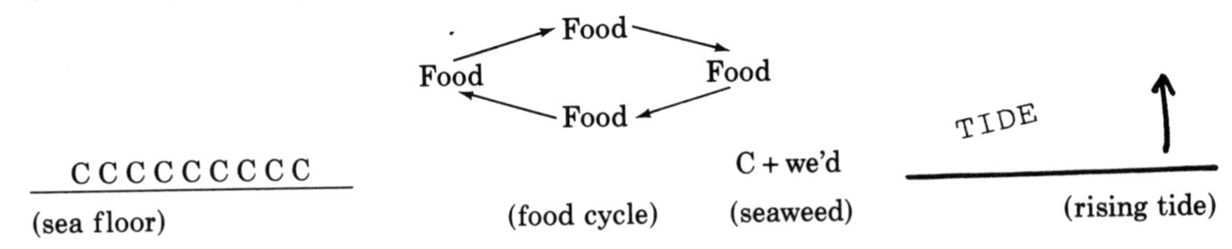


9. List five sea organisms that contain a boy's name or nickname. (Possible answers: *hal*ibut, *sal*mon, *al*gae, sm*elt*, and dia*tom*)
10. List five sea organisms that contain a girl's name or nickname. (Possible answers: *cora*l, *di*atom, s*kate*, *jelly*fish, and t*una*.)
11. Complete the puzzle.

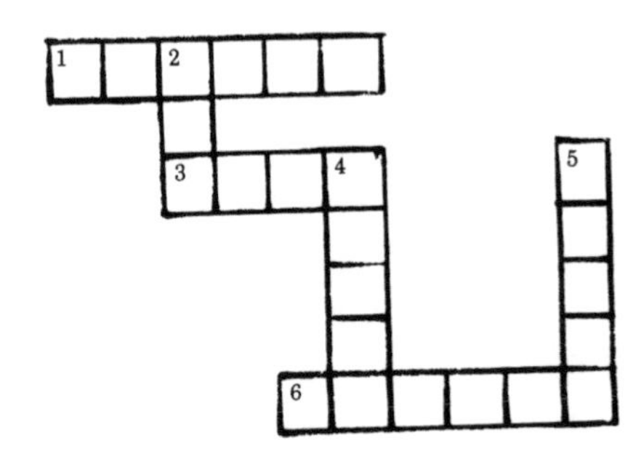


**Across**

1. A small fish (minnow)
3. A large, oily marine fish; highly prized for food (tuna)
6. The zone of swimming marine organisms (nekton)

**Down**

2. Used to catch fish (net)
4. A group of plants that includes seaweed and kelp (algae)
5. A sunken area on the sea floor (basin)

12. Expressions are fun to use, especially when they include sea life. For example, "That's a whale of an idea" or "I wish Don would clam up." Write three expressions you have heard comparing a person with a sea animal. (Possible answers: Pete's a real barracuda, John has the spine of a jellyfish, Carol acts like an old crab, Dave's skin is red like a lobster, Nan's just a shrimp, etc.)

13. A fisherman caught a five-armed starfish. He cut off each arm and threw the starfish back into the water. In time the starfish grew five new arms. If the fisherman did the same thing to three more starfish, how many arms would each starfish grow?

    (**Answer:** Unknown. Some of the starfish may grow fewer than five arms.)

14. Use the clues below to identify the sea creatures:
    - It has 10 arms but cannot write. (squid)
    - It digs through mud without a shovel. (clam)
    - It has no backbone but many spines. (sea urchin)
    - It has several arms but no biceps. (starfish)
    - A nonedible animal with a vegetable name. (sea cucumber)

# BIBLIOGRAPHY

## SECTION 2: ENERGY RESOURCES

Hoehn, Robert G. *Individualized Science Activity Cards,* "Beneath the Earth," cards 13 and 15. West Nyack, NY: The Center for Applied Research in Education, 1976.

## SECTION 3: MINERALS

Hoehn, Robert G. *Illustrated Treasury of General Science Activities,* p. 99. West Nyack, NY: Parker Publishing Company, 1975.

## SECTION 5: VOLCANOES

Hoehn, Robert G. "Ancient Hawaiian Graffiti," *Science Activities,* April/May 1987, vol. 24, no. 2, pp. 4–8.

———. *Illustrated Treasury of General Science Activities,* p. 121. West Nyack, NY: Parker Publishing Company, 1975.

## SECTION 6: EARTHQUAKES

Hoehn, Robert G. *Individualized Science Activity Cards,* "Beneath the Earth," cards 17, 18, 19, and 20. West Nyack, NY: The Center for Applied Research in Education, 1976.

Wood, Karen D., and Catherine P. Conwell. "Connecting the New with the Old," *Science Activities,* April/May 1987, vol. 24, no. 2, pp. 10–11.

## SECTION 7: FOSSILS

Hoehn, Robert G. "The Big Squeeze," *Science Activities,* April/May 1983, vol. 20, no. 2, pp. 27–30.

———. "Fossil Simulation in the Classroom," *Science Activities,* January/February 1977, vol. 14, no. 1, pp. 10–15.

———. "Fossilizing," *Science Activities,* September 1973, vol. 10, no. 1, pp. 22–34.

———. "How Plants Turn into Fossils," *Science Activities,* February/March 1982, vol. 19, no. 1, pp. 20–23.

———. "Model Fossils from Smoke and Plaster," *Science Activities* (special earth science issue), vol. 15, no. 3, pp. 26–39.

## SECTION 9: FORCES THAT SHAPE THE EARTH'S SURFACE

Hoehn, Robert G. "A Griddle Earth," *Science Activities,* December 1972, vol. 8, no. 4, pp. 14–16.

## SECTION 10: WEATHER AND CLIMATE

Hoehn, Robert G. *Illustrated Treasury of General Science Activities,* pp. 74–77. West Nyack, NY: Parker Publishing Company, 1975.

———. *Individualized Science Activity Cards,* "General Science Projects and Investigations," cards 14 and 15. West Nyack, NY: The Center for Applied Research in Education, 1976.

## SECTION 11: ASTRONOMY

Hoehn, Robert G. *Illustrated Treasury of General Science Activities,* pp. 48–60, 64–67. West Nyack, NY: Parker Publishing Company, 1975.

## SECTION 12: OCEANOGRAPHY

"Adios, Maybe, to El Niño," *Time,* September 19, 1983, vol. 122, no. 12, p. 67.

Baggett, James A. "Are America's Beaches Washing Away?" *Science World,* May 18, 1987, vol. 43, no. 18, pp. 6–9.

Hoehn, Robert G. *Illustrated Treasury of General Science Activities,* pp. 152–153, 216–217. West Nyack, NY: Parker Publishing Company, 1975.

"Rarely Seen Crustaceans Survive Climb from the Deep," *Science World,* December 15, 1986, vol. 43, no. 8, p. 8.

"Rising Seas Threaten Coastal Life," *Current Science,* May 16, 1986, vol. 7, no. 18, p. 8.

Sagan, Carl. "The Warming of the World," *Parade* magazine, February 3, 1985, pp. 9–11.

"Scientists Find Titanic and Treasure Ship," *Current Science,* November 29, 1985, vol. 71, no. 7, pp. 6–7.

"A Ten-Stop Tour of the Ocean Frontier," *Science World,* November 17, 1986, vol. 43, no. 6, pp. 18–19.

Weimer, Deborah Heiligman. "Exploring Ocean Depths—with the Help of a Robot 'Eye'," *Science World,* October 20, 1986, pp. 4–7.

Weisburd, S. "El Niño Brought the Blues, But Bliss Too," *Science News,* October 13, 1984, vol. 126, no. 15, p. 128.

# Teacher's Notes

# Teacher's Notes

# Teacher's Notes

# Teacher's Notes

# Teacher's Notes

# Teacher's Notes